Pretensions to Empire

Also by Lewis H. Lapham

The Agony of Mammon
Fortune's Child
Gag Rule
Hotel America
Imperial Masquerade
Lights, Camera, Democracy!
Money and Class in America
Theater of War
30 Satires
Waiting for the Barbarians
The Wish for Kings

Pretensions to Empire

Notes on the Criminal Folly of the Bush Administration

LEWIS H. LAPHAM

THE NEW PRESS

NEW YORK
LONDON

Requests for permission to reproduce selections from this book
should be mailed to: Permissions Department,
The New Press, 38 Greene Street, New York, NY 10013.

Published in the United States by The New Press, New York, 2006
Distributed by W. W. Norton & Company, Inc., New York

LIBRARY OF CONGRESS CATALOGING-IN-PUBLICATION DATA

Lapham, Lewis H.
    Pretensions to empire : notes on the criminal folly of the Bush administration /
Lewis H. Lapham.
        p. cm.
    Includes bibliographical references and index.
    ISBN-13: 978-1-59558-112-9
    ISBN-10: 1-59558-112-X
    1. United States—Politics and government—2001-2. United States—Foreign
relations—2001-3. Political corruption—United States. 4. Imperialism. I. Title.

E902.L368 2006
973.931—dc22

2006043330

The New Press was established in 1990 as a not-for-profit alternative to the large,
commercial publishing houses currently dominating the book publishing industry.
The New Press operates in the public interest rather than for private gain, and
is committed to publishing, in innovative ways, works of educational, cultural, and
community value that are often deemed insufficiently profitable.

www.thenewpress.com

Composition by Westchester Book Composition
This book was set in Garamond Three

Printed in the United States of America

2   4   6   8   10   9   7   5   3   1

For my son, Winston

Let the eagle soar,
Like she's never soared before
From the rocky coast to golden shore,
Let the mighty eagle soar.
Soar with healing in her wings.
As the land beneath her sings:
"Only god, no other kings."
—John Ashcroft

For my own part I wish the Bald Eagle had not been chosen the Representative of our country.... He does not get his Living honestly...he is never in good Case but like those among Men who live by Sharping and Robbing he is generally poor and often very lousy. Besides he is a rank Coward;...
—Benjamin Franklin

Contents

Preface

Suckled at the breast of the Cold War, nurtured by the fear of nuclear annihilation, fattened for half a century on the seed of profligate military spending, the notion of American world empire was a popular attraction on Washington's neoconservative think-tank circuit as long ago as 1993. The Pentagon that winter published a policy paper, "Defense Strategy for the 1990's," setting forth the doctrines of "forward deterrence," "anticipatory self-defense," and "preemptive strike" that ten years later were made manifest in the deserts of Iraq. Drafted two years previously by Dick Cheney and Colin Powell while they served as privy counselors in the first of the two Bush administrations (Cheney as Secretary of Defense, Powell as Chairman of the Joint Chiefs of Staff), the document informed the lesser nations of the earth of their inferior and subsidiary status; let any of them even begin to think of challenging the American supremacy, and America reserved the right to strangle the impudence at birth—to bomb the peasants or the palace, block the flows of oil and bank credit, change the unsanitary regime.

The ultimatum apparently was conceived for reasons having less to do with its geopolitical sense or consequences than with the need to guarantee the profits and preserve the *raison d'être* of the American military-industrial complex. The task in 1990–91 was urgent. The sudden and unlooked-for collapse of the Soviet Union had deprived the United States of an asset as precious to the national economy as General Motors and Iowa corn, the *sine qua non* that had provided nine American presidents with a just and noble cause, supplied the dark black cloth of Communist menace against which every freedom-loving politician could project the wholesome images of American innocence and goodness of heart.

A stupendous enemy, world-class and operatic, sorely missed and by no means easy to replace. But absent the threat posed by the villainous Russians, how then to maintain the Pentagon's budgets at combat strength? No more compelling question confronted Washington officialdom in 1990–91. The government had on hand an American war machine truly marvelous to behold—gun platforms of every conceivable caliber and throw weight, aircraft and naval vessels at all points of every horizon, guidance systems endowed with the wisdom of angels and armed with the judgments of doom—but other than as a means of changing lead into gold, who could say what the thing was supposed to do? Where was the tactical objective or the strategic purpose? To what end the fires of heaven, faith-based and digitally enhanced?

Pressed for a solution to the problem, Messrs. Cheney and Powell, assisted in their labors by Paul Wolfowitz (then an undersecretary of Defense, more recently the president of the World Bank) appointed America Keeper of the World's Peace, an august and solemn office requiring an appropriately impressive military staff, invincible and splendidly equipped, capable of waging, simultaneously, major wars on two continents while at the same time attending to the minor nuisances presented by terrorists here and there in the slums of the Middle East, bandits in the mountains of Afghanistan and the jungles of Colombia, mercenary armies wandering in the mists of equatorial Africa. Although fatuous in its pretension, the strategy for a new decade was conspicuously expensive and therefore a worthy answer to the patriotic call to arms, which was to defend, honor, and protect the cash flow of the nation's weapons manufacturers.

The program wasn't advertised in the display windows of the Clinton Administration, but during the years 1992–98 it gathered an increasing number of admirers in the Washington policy institutes and the neoconservative press. The consensus of the country's respectable opinion continued to shift toward the radical and reactionary right, taking its cues from Clinton's troubles with Monica Lewinsky, tuning its politics to the messianic radio talkshow frequencies and to the junkyard bombast of Newt Gingrich (R–Georgia), self-proclaimed "teacher of civilization" newly elected, in 1994, Speaker of the House of Representatives.

By the mid-1990s the conference participants at the Heritage Foundation and the American Enterprise Institute had promoted the Pentagon's sales pitch for extended subsidy into a selfless mission to save and rule the world. The gentlemen on the dais usually had served as functionaries in the administrations of Presidents Ronald Reagan and George H.W. Bush (as Secretaries of Defense or State, as Chiefs of Naval Operations or Directors of the Central Intelligence Agency), and usually they could be counted upon to preface their remarks by saying that America was always and everywhere surrounded by clever and brutal enemies. The high-end moralists in the room (sometimes Donald Rumsfeld or Richard Perle, often William Bennett accompanied by a smirk of columnists from *The Weekly Standard*) deplored the lack of virtue on the part of an American citizenry drifting in the morass of cultural decline, too easily misdirected by the lying liberal news media into the swamps of rank materialism. The after-dinner speaker invariably told the travelers from New York that the idea of government made to the measure of a provincial democratic republic (America in 1941) no longer could accommodate the interests of a global nation-state that deserved to wear the crown and name of empire (America, circa 1996).

It wasn't that anybody had intended so glorious a metamorphosis, but how could it be otherwise? The Russians had lost the Cold War, their weapons gone to rust, their economy in ruins, the statues of V. I. Lenin reduced to a rubble of broken stones. From the Chinese not even Henry Kissinger expected anything but a supply of cheap labor for another thirty years, and now that history was at an end, America embodied "the single model of human progress." The American way was the right way, and if not America, "the world's only surviving superpower," who else could lift the burden once borne on the back of imperial Rome? Reinforced by the stock market boom that carried the Dow Jones Industrial Average triumphantly across the threshold of the new millennium, the presumptions of omnipotence inflated the rhetoric to so condescending a pitch over the next five years that in March 2001, six months before the destruction of the World Trade Center, *Time* magazine gave voice to what in Washington geopolitical circles had become a matter of simple truth and common knowledge:

America is no mere international citizen. It is the dominant power in the world, more dominant than any since Rome. Accordingly, America is in the position to reshape norms, alter expectations and create new legalities. How? By unapologetic and implacable demonstrations of will.

"Tentacles of Rage," the chapter that serves as prologue to the volume in hand, published in *Harper's Magazine* in September 2004 soon after the Republican Nominating Convention in New York

City sent forth President George W. Bush to his second term in the White House, provides some of the historical context for our current state of political confusion and disgrace. The visionary sentiments floating through Madison Square Garden with the balloons were the same ones that Senator Barry Goldwater of Arizona brought into his presidential campaign in 1964, as foolish now as they were then, but during the intervening forty years, armed by the country's military-industrial genius with the power to destroy not only what remains of the American democracy but also to seize and possess as much of the world's wealth and happiness as can be fed into the mouth of never-ending war.

The other twenty-eight chapters in the book proceed in chronological sequence from August 2002 through March 2006, extending a prior sequence (also published by The New Press under the title *Theater of War*), which charts the course of the Bush Administration from its inception in the autumn of 2000 to its invasion of Iraq in the spring of 2003. Whether encountered in two volumes or one, these essays describe a march of folly, establish a record of moral incompetence and criminal intent, speak to the character of a government stupefied by its worship of money and blinded by its belief in miracles.

March 2006

PROLOGUE

Tentacles of Rage

The Republican Propaganda Mill,
a Brief History

*When, in all our history, has anyone with ideas so bizarre,
so archaic, so self-confounding, so remote from the basic
American consensus, ever got so far?*
—Richard Hofstadter

*I*n company with nearly every other historian and political journalist east of the Mississippi River in the summer of 1964, the
late Richard Hofstadter saw the Republican Party's naming of Senator Barry Goldwater as its candidate in that year's presidential election as an event comparable to the arrival of the Mongol hordes
at the gates of thirteenth-century Vienna. The "basic American consensus" at the time was firmly liberal in character and feeling, assured of a clear majority in both chambers of Congress as well as
a sympathetic audience in the print and broadcast press. Even the
National Association of Manufacturers was still aligned with the
generous impulse of Franklin Roosevelt's New Deal, accepting of
the proposition, as were the churches and the universities, that government must do for people what people cannot do for themselves.*

---

*With regard to the designation "liberal," the economist John K. Galbraith said in 1964,
"Almost everyone now so describes himself." Lionel Trilling, the literary critic, observed in*

And yet, seemingly out of nowhere and suddenly at the rostrum of the San Francisco Cow Palace in a roar of triumphant applause, here was a cowboy-hatted herald of enlightened selfishness threatening to sack the federal city of good intentions, declaring the American government the enemy of the American people, properly understood not as the guarantor of the country's freedoms but as a syndicate of quasi-communist bureaucrats poisoning the wells of commercial enterprise with "centralized planning, red tape, rules without responsibility, and regimentation without recourse." A band played "America the Beautiful," and in a high noon glare of klieg light the convention delegates beheld a militant captain of capitalist jihad ("Extremism in the defense of liberty is no vice!") known to favor the doctrines of forward deterrence and preemptive strike ("Let's lob a nuclear bomb into the men's room at the Kremlin"), believing that poverty was proof of bad character ("lazy, dole-happy people who want to feed on the fruits of somebody else's labor"), that the Democratic Party and the network news programs were under the direction of Marxist ballet dancers, that Mammon was another name for God.

The star-spangled oratory didn't draw much of a crowd on the autumn campaign trail. The electorate in 1964 wasn't interested in the threat of an apocalyptic future or the comforts of an imaginary past, and Goldwater's reactionary vision in the desert faded into the sunset of the November election won by Lyndon Johnson with 61 percent of the popular vote, the suburban sheriffs on their palomino ponies withdrawing to Scottsdale and Pasadena in the orderly and inoffensive manner of the Great Khan's horsemen retiring from the plains of medieval Europe.

1950 that "In the United States at this time, liberalism is not only the dominant but even the sole intellectual tradition." He went on to say that "there are no conservative or reactionary ideas in general circulation," merely "irritable mental gestures which seek to resemble ideas."

Departed but not disbanded. As the basic American consensus has shifted over the last thirty years from a liberal to a conservative bias, so also the senator from Arizona has come to be seen as a prophet in the western wilderness, apostle of the rich man's dream of heaven that placed Ronald Reagan in the White House in 1980 and provides the current Bush Administration with the platform on which the candidate was trundled into New York City in August for the Republican nominating convention with Arnold Schwarzenegger, the heavy law enforcement, and the paper elephants.* The speeches in Madison Square Garden affirmed the great truths now routinely preached from the pulpits of Fox News and the *Wall Street Journal*—government the problem, not the solution; the social contract a dead letter; the free market the answer to every maiden's prayer; the hollow rattle of the rhetorical brass and tin brought to mind the question that Hofstadter didn't stay to answer. How did a set of ideas both archaic and bizarre make its way into the center ring of the American political circus?

About the workings of the right-wing propaganda mills in Washington and New York I knew enough to know that the numbing of America's political senses didn't happen by mistake, but it wasn't until I met Rob Stein, formerly a senior adviser to the chairman of the

---

*The rightward movement of the country's social and political center of gravity isn't a matter of opinion or conjecture. Whether compiled by Ralph Nader or by journalists of a conservative persuasion (most recently John Micklethwait and and Adrian Wooldridge in a book entitled The Right Nation) the numbers tell the same unambiguous story—one in five Americans willing to accept identity as a liberal, one in three preferring the term "conservative"; the American public content with lower levels of government spending and higher levels of economic inequality than those pertaining in any of the Western European democracies; the United States unique among the world's developed nations in its unwillingness to provide its citizens with a decent education or fully funded health care; 40 million Americans paid less than $10 an hour, 66 percent of the population earning less than $45,000 a year; 2 million people in prison, the majority of them black and Latino; the country's largest and most profitable corporations relieved of the obligation to pay an income tax; no politician permitted to stand for public office without first professing an ardent faith in God.

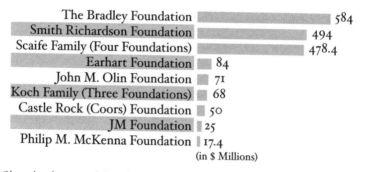

$2 BILLION ASSETS
CONSERVATIVE FOUNDATIONS (2001 ASSETS)

The Bradley Foundation	584
Smith Richardson Foundation	494
Scaife Family (Four Foundations)	478.4
Earhart Foundation	84
John M. Olin Foundation	71
Koch Family (Three Foundations)	68
Castle Rock (Coors) Foundation	50
JM Foundation	25
Philip M. McKenna Foundation	17.4

(in $ Millions)

Charts based on research by Rob Stein.

Democratic National Committee, that I came to fully appreciate the nature and the extent of the re-education program undertaken in the early 1970s by a cadre of ultraconservative and self-mythologizing millionaires bent on rescuing the country from the hideous grasp of Satanic liberalism. To a small group of Democratic activists meeting in New York City in late February, Stein had brought thirty-eight charts diagramming the organizational structure of the Republican "Message Machine," an octopus-like network of open and hidden microphones that he described as "perhaps the most potent, independent institutionalized apparatus ever assembled in a democracy to promote one belief system."

It was an impressive presentation, in large part because Stein didn't refer to anybody as a villain, never mentioned the word "conspiracy." A lawyer who also managed a private equity investment fund—i.e., a man unintimidated by spreadsheets and indifferent to the seductions of the pious left—Stein didn't begrudge the manufacturers of corporatist agitprop the successful distribution of their product in the national markets for the portentous catch-phrase and the camera-ready slogan. Having devoted several months to his search through the available documents, he was content to let the facts speak for themselves—fifty funding agencies of different dimensions and varying degrees of ideological fervor, nominally philanthropic but zealous in their common hatred of the liberal enemy, disbursing the collective sum of roughly $3 billion over a period of thirty years for the fabrication of "irritable mental gestures which seek to resemble ideas."

The effort had taken many forms—the publication of expensively purchased and cleverly promoted tracts (Milton Friedman's *Free to Choose,* Charles Murray's *Losing Ground,* Samuel Huntington's *The Clash of Civilizations*), a steady flow of newsletters from more than 100 captive printing presses (among them those at The Heritage Foundation, Accuracy in the Media, the American Enterprise Institute and the Center for the Study of Popular Culture), generous distributions of academic programs and visiting professorships (to Harvard, Yale, and Stanford universities), the passing along of sound-bite slanders (to Bill O'Reilly and Matt Drudge), the formulation of newspaper op-ed pieces (for the *San Antonio Light* and the *Pittsburgh Post-Gazette* as well as for the *Sacramento Bee* and the *Washington Times*). The prolonged siege of words had proved so successful in its result that on nearly every question of foreign or domestic policy in this year's presidential campaign, the frame and terms of the debate might as well have been assembled in Taiwan by Chinese child labor working from patterns furnished by the authors of ExxonMobil's annual report.

No small task and no mean feat, and as I watched Stein's diagrams take detailed form on a computer screen (the directorates of the Leadership Institute and Capital Research Center all but identical with that of The Philanthropy Roundtable, Richard Mellon Scaife's money dispatched to the Federalist Society as well as to *The American Spectator*), I was surprised to see so many familiar names—publications to which I'd contributed articles, individuals with whom I was acquainted—and I understood that Stein's story was one that I could corroborate, not with supplementary charts or footnotes but on the evidence of my own memory and observation.

The provenience of the Message Machine Stein traced to the recognition in the part of the country's corporate gentry in the late 1960s that they lacked the intellectual means to comprehend, much less quell or combat, the social and political turmoil then engulfing the whole of American society, and if I had missed Goldwater's foretelling of an apocalyptic future in the Cow Palace, I remembered my own encounter with the fear and trembling of what was still known as "The Establishment," four years later and 100 miles to the north at the July encampment of San Francisco's Bohemian Club. Over a period of three weeks every summer, the 600-odd members of the club, most of them expensive ornaments of the American haute bourgeosie, invite an equal number of similarly fortunate guests to spend as many days as their corporate calendars permit within a grove of handsome redwood trees, there to listen to the birdsong, interest one another in various business opportunities, exchange misgivings about the restlessness of the deutschmark and the yen.

In the summer of 1968 the misgivings were indistinguishable from panic. Martin Luther King had been assassinated; so had Robert Kennedy, and everywhere that anybody looked the country's institutional infrastructure, also its laws, customs, best-loved truths, and fairy tales, seemed to be collapsing into anarchy and chaos—black people rioting in the streets of Los Angeles and Detroit, American soldiers killing their officers in Vietnam, long-haired hippies stoned on drugs or drowned in the bathtubs of Bel Air, shorthaired feminists playing with explosives instead of dolls, the Scottsdale and Pasadena sheriffs' posses preparing their palomino ponies to stand firm in the face of an urban mob.

Historians revisiting in tranquillity the alarums and excursions of the Age of Aquarius know that Revolution Now was neither imminent nor likely—the economy was too prosperous, the violent gestures of rebellion contained within too small a demographic, mostly rich kids who could afford the flowers and the go-go boots—but in the hearts of the corporate chieftains wandering among the redwood trees in the Bohemian Grove in July 1968, the fear was palpable and genuine. The croquet lawn seemed to be sliding away beneath their feet, and although they knew they were in trouble, they didn't know why. Ideas apparently mattered, and words were maybe more important than they had guessed; unfortunately, they didn't have any. The American property-holding classes tend to be embarrassingly ill at ease with concepts that don't translate promptly into money, and the beacons of conservative light shining through the liberal fog of the late 1960s didn't come up to the number of clubs in Arnold Palmer's golf bag. The company of the commercial faithful gathered on the banks of California's Russian River could look for succor to Goldwater's autobiography, *The Conscience of a Conservative,* to William F. Buckley's editorials in *National Review,* to the novels of Ayn Rand. Otherwise they were as helpless as unarmed sheepherders surrounded by a Comanche war party on the old Oklahoma frontier before the coming of the railroad and the six-gun.

The hope of their salvation found its voice in a 5,000-word manifesto written by Lewis Powell, a Richmond corporation lawyer, and circulated in August 1971 by the United States Chamber of Commerce under the heading *Confidential Memorandum; Attack on the American Free Enterprise System.* Soon to be appointed to the Supreme Court, lawyer Powell was a man well-known and much respected by the country's business community; within the legal profession he was regarded as a prophet. His heavy word of warning fell upon the legions of reaction with the force of Holy Scripture: "Survival of what we call the free enterprise system," he said, "lies in organization, in careful long-range planning and implementation, in consistency of action over an indefinite period of years, in the scale of financing available only through joint effort, and in the political power available only through united action and national organizations."

The venture capital for the task at hand was provided by a small sewing circle of rich philanthropists—Richard Mellon Scaife in Pittsburgh, Lynde and Harry Bradley in Milwaukee, John Olin in New York City, the Smith Richardson family in North Carolina, Joseph Coors in Denver, David and Charles Koch in Wichita—who entertained visions of an America restored to the safety of its mythological past—small towns like those seen in prints by Currier and Ives, cheerful factory workers whistling while they worked, politicians as wise as Abraham Lincoln and as brave as Teddy Roosevelt, benevolent millionaires presenting Christmas turkeys to deserving elevator operators, the sins of the flesh deported to Mexico or France. Suspicious of any fact that they hadn't known before the age of six, the wealthy saviors of the Republic also possessed large reserves of paranoia, and if the world was going rapidly to rot (as any fool could plainly see) the fault was to be found in everything and anything tainted with a stamp of liberal origin—the news media and the universities, income taxes, Warren Beatty, transfer payments to the undeserving poor, restraints of trade, Jane Fonda, low interest rates, civil liberties for unappreciative minorities, movies made in Poland, public schools.*

Although small in comparison with the sums distributed by the Ford and Rockefeller foundations, the money was ideologically sound, and it was put to work leveraging additional contributions (from corporations as well as from other like-minded foundations),

---

*The various philanthropic foundations under the control of the six families possess assets estimated in 2001 at $1.7 billion. Harry Bradley was an early and enthusiastic member of the John Birch Society; Koch Industries in the winter of 2000 agreed to pay $30 million (the largest civil fine ever imposed on a private American company under any federal environmental law) to settle claims related to 300 oil spills from its pipelines in six states.

acquiring radio stations, newspapers, and journals of opinion, bankrolling intellectual sweatshops for the making of political and socioeconomic theory. Joseph Coors established The Heritage Foundation with an initial gift of $250,000 in 1973, the sum augmented over the next few years with $900,000 from Richard Scaife; the American Enterprise Institute was revived and fortified in the late seventies with $6 million from the Howard Pew Freedom Trust; the Cato Institute was set up by the Koch family in 1977 with a gift of $500,000. If in 1971 the friends of American free enterprise could turn for comfort to no more than seven not very competent sources of inspiration, by the end of the decade they could look to eight additional installations committed to "joint effort" and "united action."* The senior officers of the Fortune 500 companies meanwhile organized the Business Roundtable, providing it by 1979 with a rich endowment for the hiring of resident scholars loyal in their opposition to the tax and antitrust laws.

The quickening construction of Santa's workshops outside the walls of government and the academy resulted in the increased production of pamphlets, histories, monographs, and background briefings intended to bring about the ruin of the liberal idea in all of its institutionalized forms—the demonization of the liberal press, the disparagement of liberal sentiment, the destruction of liberal education—and by the time Ronald Reagan arrived in triumph at the White House in 1980 the assembly lines were operating at full capacity. Well in advance of inauguration day the Christmas elves had churned out so much paper that had they been told to do so, they could have shredded it into tickertape and welcomed the new cowboy-hatted herald of enlightened selfishness with a parade like none other ever before seen by man or beast.

---

*Paul Weyrich, the first director of The Heritage Foundation, and often described by his admirers as "the Lenin of social conservatism," seldom was at a loss for a military analogy: "If your enemy has weapons systems working and is killing you with them, you'd better have weapons systems of your own."

Unshredded, the paper was the stuff of dreams from which was made *Mandate for Leadership,* the "bible" presented by The Heritage Foundation to Mr. Reagan in the first days of his presidency with the thought that he might want to follow its architectural design for an America free at last from "the tyranny of the Left," rescued from the dungeons of "liberal fascism," once again a theme park built by nonunion labor along the lines of Walt Disney's gardens of synthetic Eden.

Signs of the newly minted intellectual dispensation began showing up in the offices of *Harper's Magazine* in 1973, the manuscripts invariably taking the form of critiques of one or another of the absurdities then making an appearance before the Washington congressional committees or touring the New York literary scene with Susan Sontag and Norman Mailer. Over a period of several years the magazine published articles and essays by authors later to become well-known apologists for the conservative creed, among them George Gilder, Michael Novak, William Tucker, and Philip Terzian; if their writing in the early seventies was remarkable both for its clarity and wit, it was because they chose topics of opportunity that were easy to find and hard to miss.

The liberal consensus hadn't survived the loss of the Vietnam War. The subsequently sharp reduction of the country's moral and economic resources was made grimly apparent by the impeachment of Richard Nixon and the price of Arab oil, and it came to be understood that Roosevelt's New Deal was no longer on offer. Acting on generous impulse and sustained by the presumption of limitless wealth, the American people had enacted legislation reflecting their best hopes for racial equality and social justice (a.k.a. Lyndon Johnson's "Great Society"), but any further efforts at transformation clearly were going to cost a great deal more money than the voters

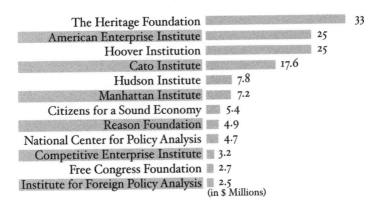

NATIONAL "THINK TANKS" (2001 BUDGETS)

The Heritage Foundation	33
American Enterprise Institute	25
Hoover Institution	25
Cato Institute	17.6
Hudson Institute	7.8
Manhattan Institute	7.2
Citizens for a Sound Economy	5.4
Reason Foundation	4.9
National Center for Policy Analysis	4.7
Competitive Enterprise Institute	3.2
Free Congress Foundation	2.7
Institute for Foreign Policy Analysis	2.5

(in $ Millions)

were prepared to spend. Also a good deal more thought than the country's liberal-minded intelligentsia were willing to attempt or eager to provide. The universities chose to amuse themselves with the crossword puzzles of French literary theory, and in the New York media salons the standard-bearers of America's political conscience were content to rest upon what they took to be their laurels, getting by with the striking of noble poses (as friends of the earth or the Dalai Lama) and the expression of worthy emotions (on behalf of persecuted fur-seals and oppressed women). The energies once contained within the nucleus of a potent idea escaped into the excitements of the style incorporated under the rubrics of Radical Chic, and the messengers bringing the good news of conservative reaction moved their gospel-singing tent show into an all but deserted public square.

Their chief talents were those of the pedant and the critic, not those of the creative imagination, but they well understood the art of merchandising and the science of cross-promotion, and in the middle 1970s anybody wishing to appreciate the character and purpose of the emerging conservative putsch could find no better informant than Irving Kristol, then a leading columnist for the *Wall Street Journal,* the author of well-received books (*On the Democratic Idea in America* and *Two Cheers for Capitalism*), trusted counselor and adjunct sage at the annual meetings of the Business Roundtable. As a youth in the late 1930s, at a time when literary name and reputation accrued to the accounts of the *soi-disant* revolutionary left, Kristol had proclaimed himself a disciple of Leon Trotsky, but then the times changed, the winds of fortune shifting from east to west, and after a stint as a CIA asset in the 1950s, he had carried his pens and papers into winter quarters on the comfortably upholstered bourgeois right.

On first meeting the gentleman at a literary dinner in New York's Century Club, I remember that I was as much taken by the ease and grace of his manner as I was impressed by his obvious intelligence. A man blessed with a sense of humor, his temperament and tone of mind more nearly resembling that of a sophisticated dealer in art and antiques than that of an academic scold, he praised *Harper's Magazine* for its publication of Tom Wolfe's satirical pieces, also for the prominence that it had given to the essays of Senator Daniel Patrick Moynihan, and I was flattered by his inclination to regard me as an editor-of-promise who might be recruited to the conservative cause, presumably as an agent in place behind enemy lines. The American system of free enterprise, he said, was being attacked by the very people whom it most enriched—i.e., by the pampered children of privilege disturbing the peace of the Ivy League universities, doing lines of cocaine in Manhattan discotheques, making decadent movies in Hollywood—and the time had come to put an end to their dangerous and self-indulgent nonsense. Nobody under the age of thirty knew what anything cost, and even the senior faculty at Princeton had forgotten that it was none other than the great Winston Churchill who had said, "Cultured people are merely the glittering scum which floats upon the deep river of production."

In the course of our introductory conversation Kristol not only referred me to other old masters whom I might wish to re-read (among them Plutarch, Gibbon, and Edmund Burke); he also explained something of his technique as an intellectual entrepreneur. Despite the warning cries raised by a few prescient millionaires far from the fashionable strongholds of the effeminate east, the full membership of the American oligarchy still wasn't alive to the threat of cultural insurrection, and in order to awaken the management to a proper sense of its dire peril, Kristol had been traveling the circuit of the country's corporate boardrooms, soliciting contributions given in memory of Friedrich von Hayek, encouraging the automobile companies to withdraw their advertising budgets from any media outlet that declined to echo their social and political prejudices.

"Why empower your enemies?" he said. "Why throw pearls to swine?"*

Although I didn't accept Kristol's invitation to what he called the "intellectual counterrevolution," I often ran across him during the next few years at various symposia addressed to the collapse of the nation's moral values, and I never failed to enjoy his company or his conversation. Among all the propagandists pointing out the conservative path to glory, Kristol seemed to me the brightest and the best, and I don't wonder that he eventually became one of the four or five principal shop stewards overseeing the labors of the Republican message machine.

---

*Henry Ford II expressed a similar thought on resigning as a trustee of the Ford Foundation in late 1976. Giving vent to his confusion, annoyance, and dismay, he took the trouble to write a letter to the staff of the foundation reminding them that they were associated with "a creature of capitalism." Conceding that the word might seem "shocking" to many of the people employed in the vineyards of philanthropy, Mr. Ford proceeded to his defense of the old ways and old order:

"I'm not playing the role of the hard-headed tycoon who thinks all philanthropoids are Socialists and all university professors are Communists. I'm just suggesting to the trustees and the staff that the system that makes the foundation possible very probably is worth preserving."

It was at Kristol's suggestion that I met a number of the fund-raising people associated with the conservative program of political correctness, among them Michael Joyce, executive director in the late seventies of the Olin Foundation. We once traveled together on a plane returning to New York from a conference that Joyce had organized for a college in Michigan, and somewhere over eastern Ohio he asked whether I might want to edit a new journal of cultural opinion meant to rebut and confound the ravings of *The New York Review of Books*. The proposition wasn't one in which I was interested, but the terms of the offer—a handsome annual salary to be paid for life even in the event of my resignation or early retirement—spoke to the seriousness of the rightist intent to corner and control the national market in ideas.*

The work went more smoothly as soon as the Reagan Administration had settled itself in Washington around the fountains and reflecting pools of federal patronage. Another nine right-thinking foundations established offices within a short distance of Capitol Hill or the Hay-Adams hotel (most prominent among them the Federalist Society and the Center for Individual Rights); more corporations sent more money; prices improved for ideological piece-work (as much as $100,000 a year for some of the brand-name scholars at Heritage and AEI), and eager converts to the various sects of the conservative faith were as thick upon the ground as maple leaves in autumn. By the end of Reagan's second term the propaganda mills were spending $100 million a year on the manufacture and sale of their product, invigorated by the sense that once again it was morning in America and redoubling their efforts to transform their large store of irritable mental gestures into brightly

The proposed journal appeared in 1982 as The New Criterion, *promoted as a "staunch defender" of high culture, "an articulate scourge of artistic mediocrity and intellectual mendacity wherever they are found." Joyce later took over direction of the Bradley Foundation, where he proved to be as deft as Weyrich and Kristol at what the movement conservatives liked to call the wondrous alchemy of turning intellect into influence.*

packaged policy objectives—tort reform, school vouchers, less government, lower taxes, elimination of the labor unions, bigger military budgets, higher interest rates, reduced environmental regulation, privatization of social security, downsized Medicaid and Medicare, more prisons, better surveillance, stricter law enforcement.

If production increased at a more handsome pace than might have been dreamed of by Richard Scaife or hoped for by Irving Kristol, it was because the project had been blessed by Almighty God. The Christian right had come into the corporate fold in the late 1970s. Abandoning the alliance formed with the conscience of the liberal left during the Great Depression (the years of sorrow and travail when money was not yet another name for Jesus), the merchants of spiritual salvation had come to see that their interests coincided with those of the insurance companies and the banks. The American equestrian classes were welcome to believe that slack-jawed dope addicts had fomented the cultural insurrection of the 1960s; Jerry Falwell knew that it had been the work of Satan, Satan himself and not one of his students at the University of California, who had loosed a plague of guitarists upon the land, tempted the news media to the broadcast of continuous footage from Sodom and Gomorrah, impregnated the schools with indecent interpretations of the Bible, which then gave birth to the monster of multiculturalism that devoured the arts of learning. Together with Paul Weyrich at The Heritage Foundation, Falwell sponsored the formation of the Moral Majority in 1979, at about the same time and in much the same spirit that Pat Robertson, the Christian televangelist, sent his congregation a fund-raising letter saying that feminists encourage women to "leave their husbands, kill their children, practice witchcraft, destroy capitalism and become lesbians." Before Ronald Reagan was elected to a second term the city of God signed a nonaggression pact with the temple of Mammon, their combined forces waging what came to be known as "The Culture War."

MASS MEDIA DISTRIBUTION

$300M CONSERVATIVE MESSAGE MACHINE

TELEVISION
Pat Robertson's *700 Club*
Fox News Channel
MSNBC's *Scarborough Country*
Oliver North's *War Stories*

RADIO
The Rush Limbaugh Show
The Cal Thomas Commentary
Radio America

PUBLISHING
Eagle Publishing, Inc.

NEWSPAPERS
The Washington Times
The Wall Street Journal

WEBSITES
Townhall.com
AnnCoulter.com

The Cold War against the Russians was fading into safe and nostalgic memory, and the tellers of the great American fairy tale (the one about the precious paradise ever in need of an invincible defense) found themselves in pressing need of other antagonists to take the place of the grim and harmless ogre in the northern snow. The Japanese couldn't play the part because they were lending the United States too much money; the Colombian drug lords were too few and too well connected in Miami; Manuel Noriega failed the audition; the Arab oil cartel was broke; and the Chinese were busy making shirts for Ralph Lauren.

In the absence of enemies abroad, the protectors of the American dream at home began looking for domestic signs of moral weakness rather than foreign shows of military strength; instead of examining the dossiers of distant tyrants, they searched the local newspapers for flaws in the American character, and the surveillance satellites over Leipzig and Sevastopol were reassigned stations over metropolitan Detroit and the Hollywood studios filming *Dynasty* and *Dallas*. Within a matter of months the conservative committees of public safety rounded up as suspects a motley crowd of specific individuals and general categories of subversive behavior and opinion—black male adolescents as well as elderly female Buddhists, the *New York Times,* multiculturalists of all descriptions, the 1960s, welfare mothers, homosexuals, drug criminals, illegal immigrants, performance artists. Some enemies of the state were easier to identify than others, but in all instances the reactionary tellers of the tale relied on images seen in dreams or Arnold Schwarzenegger movies rather than on the lessons of their own experience.

For a few years I continued to attend convocations sponsored by the steadily proliferating agencies of the messianic right, but although the discussions were held in increasingly opulent settings—the hotel accommodations more luxurious, better food, views of the mountains as well as the sea—by 1985 I could no longer stomach either the sanctimony or the cant. With the coming to power of the Reagan Administration most of the people on the podium or the tennis court were safely enclosed within the perimeters of orthodox opinion and government largesse, and yet they persisted in casting themselves as rebels against "the system," revolutionary idealists being hunted down like dogs by a vicious and still active liberal prosecution. The pose was as ludicrous as it was false. The leftist impulse had been dead for ten years, ever since the right-wing Democrats in Congress had sold out the liberal portfolio of President Jimmy Carter and revised the campaign-finance laws to suit the convenience of their corporate patrons. Nor did the news media present an obstacle. By 1985 the *Wall Street Journal* had become the newspaper of record most widely read by the people who made the decisions about the country's economic policy; the leading editorialists in the *New York Times* (A. M. Rosenthal, William Safire) as well as in the *Washington Post* (George Will, Richard Harwood, Meg Greenfield) ably defended the interests of the status quo; the vast bulk of the nation's radio talk shows (reaching roughly 80 percent of the audience) reflected a conservative bias, as did all but one or two of the television talk shows permitted to engage political topics on PBS. In the pages of the smaller journals of opinion (*National Review, Commentary, The American Spectator, The National Interest, The New Criterion, The Public Interest, Policy Review,* etc.) the intellectual décor, much of it paid for by the Olin and Scaife foundations, was matched to the late-Victorian tastes of Rudyard Kipling and J. P. Morgan. The voices of conscience that attracted the biggest crowds on the nation's lecture circuit were those that spoke for one or another of the parties of the right, and together with the chorus of religious broadcasts and pamphlets (among them Pat Robertson's *700 Club* and the publications under the direction of Jerry Falwell and the Reverend Sun Myung Moon), they enveloped the country in an all but continuous din of stereophonic, right-wing sound.

The facts seldom intruded upon the meditations of the company seated poolside at the conferences and symposia convened to bemoan America's fall from grace, and I found it increasingly depressing to listen to prerecorded truths dribble from the mouths of writers once willing to risk the chance of thinking for themselves. Having exchanged intellectual curiosity for ideological certainty, they had forfeited their powers of observation as well as their senses of humor; no longer courageous enough to concede the possibility of error or enjoy the play of the imagination, they took an interest only in those ideas that could be made to bear the weight of solemn doctrine, and they cried up the horrors of the culture war because their employers needed an alibi for the disappearances of the country's civil liberties and a screen behind which to hide the privatization (a.k.a. the theft) of its common property—the broadcast spectrum as well as the timber, the water, and the air, the reserves of knowledge together with the mineral deposits and the laws. Sell the suckers on the notion that their "values" are at risk (abortionists escaping the nets of the Massachusetts state police, pornographers and cosmetic surgeons busily at work in Los Angeles, farm families everywhere in the Middle West becoming chattels of the welfare state) and maybe they won't notice that their pockets have been picked.

So many saviors of the republic were raising the alarm of culture war in the middle eighties that I now can't remember whether it was Bob Bartley writing in the *Wall Street Journal* or William Bennett speaking from his podium at the National Endowment for the Humanities who said that at Yale University the students were wallowing in the joys of sex, drugs, and Karl Marx, disporting themselves on the New Haven green in the reckless manner of nymphs and satyrs on a Grecian urn. I do remember that at one of the high-end policy institutes in Manhattan I heard the tale told by Norman Podhoretz, then the editor at *Commentary,* the author of several contentious books (*Making It* and *Why We Were in Vietnam*), and a rabid propagandist for all things anti-liberal. What he had to say about Yale was absurd, which I happened to know because that same season I was teaching a seminar at the college. More than half the number of that year's graduating seniors had applied for work at the First Boston Corporation, and most of the students whom I'd had the chance to meet were so busy finding their way around the Monopoly board of the standard American success (figuring the angles of approach to business school, adding to the network of contacts in their Filofaxes) that they didn't have the time to waste on sexual digressions either literal or figurative. When I attempted to explain the circumstance to Podhoretz, he wouldn't hear of it. Not only was I misinformed, I was a liberal and therefore both a liar and a fool. He hadn't been in New Haven in twenty years, but he'd read William F. Buckley's book (*God and Man at Yale,* published in 1951), and he knew (because the judgment had been confirmed by something he'd been told by Donald Kagan in 1978) that the college was a sinkhole of depraved sophism. He knew it for a fact, knew it in the same way that Jerry Falwell knew that it was Satan who taught Barbra Streisand how to sing.

If Kristol was the most engaging of the agents provocateur whom I'd encountered on the conservative lecture circuit in the 1980s, Podhoretz was the dreariest—an apparatchik in the old Soviet sense of the word who believed everything he wished to prove and could prove everything he wished to believe, bringing his patrons whichever words might serve or please, anxious to secure a place near or at the boot of power. Unfortunately it was Podhoretz, not Kristol, who exemplified the character and tone of mind that edged the American conservative consensus ever further to the right during the decade of the 1990s.

The networks of reactionary opinion once again increased their rates of production, several additional foundations recruited to the cause, numerous activist organizations coming on line, together with new and improved media outlets (most notably Rupert Murdoch's Fox News and *Weekly Standard*) broadcasting the gospels according to saints Warren Harding and William McKinley. By 1994 the Conservative Political Action Conference was attracting as many as 4,000 people, half of them college students, to its annual weekend in Arlington, Virginia, there to listen to the heroes of the hour (G. Gordon Liddy, Ralph Reed, Oliver North) speak from stages wrapped in American flags. Americans for Tax Reform under

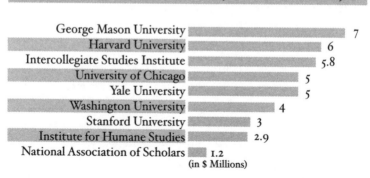

STUDENTS AND SCHOLARS (2001 ESTIMATES)

George Mason University — 7
Harvard University — 6
Intercollegiate Studies Institute — 5.8
University of Chicago — 5
Yale University — 5
Washington University — 4
Stanford University — 3
Institute for Humane Studies — 2.9
National Association of Scholars — 1.2
(in $ Millions)

the direction of Grover Norquist declared its intention to shrink the federal government to a size small enough "to drown," like one of the long-lost hippies in Bel Air, "in a bathtub."

Although as comfortably at home on Capitol Hill as in the lobbies of the corporate law firms on K Street, and despite their having learned to suck like newborn lambs at the teats of government patronage (Kristol's son, William, serving as public-relations director to Vice President Dan Quayle; Podhoretz's son-in-law, Elliot Abrams, a highly placed official within the Reagan Administration subsequently indicted for criminal misconduct), the apologists for the conservative cause continued to pose as embattled revolutionaries at odds with the "Tyranny of the Left." The pretense guaranteed a steady flow of money from their corporate sponsors, and the unexpected election of Bill Clinton in 1992 offered them yet another chance to stab the corpse of the liberal Goliath. The smearing of the new president's name and reputation began as soon as he committed the crime of entering the White House. *The American Spectator,* a monthly journal financed by Richard Scaife, sent its scouts west into Arkansas to look for traces of Clinton's semen on the pine trees and the bar stools. It wasn't long before Special Prosecutor Kenneth Starr undertook his obsessive inspection of the president's bank records, soul, and penis. Summoning witnesses with the fury of a suburban Savonarola, Starr set forth on an exploration of the Ozark Mountains, questioning the natives about wooden Indians and painted women. For four years he camped in the wilderness, and even after he was allowed to examine Monica Lewinsky's lingerie drawer, his search for the weapon of mass destruction proved as futile as the one more recently conducted in Iraq.

Although unable to match Starr's prim self-righteousness, Newt Gingrich, the Republican congressman from Georgia elected speaker of the House in 1995, presented himself as another champion of virtue (a self-proclaimed "Teacher of the Rules of Civilization") willing to lead the American people out of the desolation of a liberal

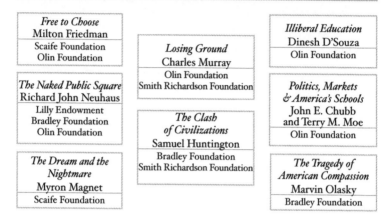

**EIGHT INFLUENTIAL BOOKS AND
THE FOUNDATIONS WHO SPONSORED THEM**

Free to Choose Milton Friedman Scaife Foundation Olin Foundation		*Illiberal Education* Dinesh D'Souza Olin Foundation
	Losing Ground Charles Murray Olin Foundation Smith Richardson Foundation	
The Naked Public Square Richard John Neuhaus Lilly Endowment Bradley Foundation Olin Foundation		*Politics, Markets & America's Schools* John E. Chubb and Terry M. Moe Olin Foundation
	The Clash of Civilizations Samuel Huntington Bradley Foundation Smith Richardson Foundation	
The Dream and the Nightmare Myron Magnet Scaife Foundation		*The Tragedy of American Compassion* Marvin Olasky Bradley Foundation

wasteland. Like Starr and Podhoretz (also like the newscasters who now decorate the right-wing television studios), Gingrich had a talent for bearing grudges. During his sixteen years in Congress he had acquired a reputation (not undeserved) for being nasty, brutish, and short, eventually coming to stand as the shared and shining symbol of resentment that bound together the several parties of the disaffected right—the Catholic conservatives with the Jewish neoconservatives, the libertarians with the authoritarians, the evangelical nationalists with the paranoid monetarists, Pat Robertson's Christian Coalition with the friends of the Ku Klux Klan. Within a few months of his elevation to the speaker's chair, Gingrich bestowed on his fellow-plaintiffs his Contract with America, a plan for rooting out the last vestiges of liberal heresy in the mind of government. As mean-spirited in its particulars as the *Mandate for Leadership* handed to Ronald Reagan in 1980, the contract didn't become law, but it has since provided the terms of enlightened selfishness that shape and inspire the policies of the current Bush Administration.

During the course of the 1990s I did my best to keep up with the various lines of grievance developing within the several sects of the conservative remonstrance, but although I probably read as many as 2,000 presumably holy texts (Peggy Noonan's newspaper editorials and David Gelernter's magazine articles as well as the soliloquies of Rush Limbaugh and the sermons of Robert Bork), I never learned how to make sense of the weird and too numerous inward contradictions. How does one reconcile the demand for small government with the desire for an imperial army, apply the phrases "personal initiative" and "self-reliance" to corporation presidents utterly dependent on the federal subsidies to the banking, communications, and weapons industries, square the talk of "civility" with the strong-arm methods of Kenneth Starr and Tom DeLay, match the warmhearted currencies of "conservative compassion" with the cold cruelty of "the unfettered free market," know that human life must be saved from abortionists in Boston but not from cruise missiles in Baghdad? In the glut of paper I could find no unifying or fundamental principle except a certain belief that money was good for rich people and bad for poor people. It was the only point on which all the authorities agreed, and no matter where the words were coming from (a report on federal housing, an essay on the payment of Social Security, articles on the sorrow of the slums or the wonder of the U.S. Navy) the authors invariably found the same abiding lesson in the tale—money ennobles rich people, making them strong as well as wise; money corrupts poor people, making them stupid as well as weak.

But if a set of coherent ideas was hard to find in all the sermons from the mount, what was not hard to find was the common tendency to believe in some form of transcendent truth. A religious as opposed to a secular way of thinking. Good versus Evil, right or wrong, saved or damned, with us or against us, and no light-minded trifling with doubt or ambiguity. Or, more plainly and as a young disciple of Ludwig Von Mises had said, long ago in the

1980s in one of the hospitality tents set up to welcome the conservative awakening to a conference on a beach at Hilton Head, "Our people deal in absolutes."

Just so, and more's the pity. In place of intelligence, which might tempt them to consort with wicked or insulting questions for which they don't already possess the answers, the parties of the right substitute ideology, which, although sometimes archaic and bizarre, is always virtuous.

Virtuous, but not necessarily the best means available to the running of a railroad or a war. The debacle in Iraq, like the deliberate impoverishment of the American middle class, bears witness to the shoddiness of the intellectual infrastructure on which a once democratic republic has come to stand. Morality deemed more precious than liberty; faith-based policies and initiatives ordained superior to common sense.

As long ago as 1964 even William F. Buckley understood that the thunder on the conservative right amounted to little else except the sound and fury of middle-aged infants banging silver spoons, demanding to know why they didn't have more—more toys, more time, more soup; when Buckley was asked that year what the country could expect if it so happened that Goldwater was elected president, he said, "That might be a serious problem." So it has proved, if not under the baton of the senator from Arizona then under the direction of his ideologically correct heirs and assigns. An opinion poll taken in 1964 showed 62 percent of the respondents trusting the government to do the right thing; by 1994 the number had dwindled to 19 percent. The measure can be taken as a tribute to the success of the Republican propaganda mill that for the last forty years has been grinding out the news that all government is bad, and that the word "public," in all its uses and declensions (public service, citizenship, public health, community, public park, commonwealth, public school, etc.), connotes inefficiency and waste. The dumbing down of the public discourse follows as the day the night, and so it comes as no surprise that both candidates in this year's presidential election present themselves as handsome images consistent with those seen in Norman Rockwell's paintings or the prints of Currier and Ives, as embodiments of what they call "values" rather than as the proponents of an idea.

September 2004

Power Points

People never lie so much as after a hunt, during a war or before an election.
—Otto von Bismarck

*I*f the waging of the war on terrorism could consist of nothing else except a savage shuffling of paper, the Bush Administration undoubtedly would win it in a week. The CIA might not know where to find Osama bin Laden or Mullah Mohammad Omar; the FBI might lose track of two murderous Al Qaeda operatives who listed their address in the San Diego phone book while learning how to fly a plane into the World Trade Center; but when it comes to the tactical deployment of a flowchart, the mobilization of new investigative guidelines, or the strategic reenforcement of a press statement, our field commanders at the microphones in Washington show themselves the equal of Napoleon.

Although the meaning of the words tends to get lost in the barrage of patriotic slogan that since September 11 has covered with the smoke of victory the landscapes of defeat, the confusion proved useful to the news conference on May 30 at which the Justice Department announced, for the third time in almost as many months, yet another reorganization of the FBI. Attorney General John Ashcroft executed the maneuver with his customary audacity, but he supported it with so many deft flanking movements (among them the concealment of his plan from the Congress) that it wasn't easy to follow the line of his advance or to know who he thought was the enemy. Two weeks after he had come and gone in the vivid glare of the Washington television cameras, I found that in New York I was still talking to people who didn't know what he had said. They had studied the transcript of his remarks, also the fifty-odd pages of supplementary power points, but they were having trouble with the translations into English. To take Ashcroft's statement at face value was to be confronted by an impregnable salient of opaque abstraction; to read his text as a not-so-subtle form of sarcasm was to admit that the administration of the country's laws had fallen into the hands of scoundrels instead of fools. Neither interpretation inspired trust in government during what was being touted as a "time of war," which is why I was asked for an editor's opinion, and how I came to read the Justice Department's Fact Sheet entitled "Crafting an Overall Blueprint for Change, Reshaping the FBI's Priorities." The brief summary itemizes the Bureau's newly regrouped purposes as well as the ones already installed by last October's passage of the U.S.A. Patriot Act. As was to be expected, the language lent itself to more than one way of connecting the dots; also as expected, certain phrases didn't suggest obvious English equivalents ("Refocus our resources on frontline positions"; "Measure accountability through outcomes and results, not by inputs"), but I could make enough sense of the less obscure imperatives to appreciate the Bureau's rededication to the highest principles of participatory fascism.

The phrases in small caps appear as given in the Fact Sheet; I've rearranged the sequence to reflect what I take to be the line of argument:

PROTECT THE UNITED STATES FROM TERRORIST ATTACK

A praiseworthy objective, but, by the government's own admission, unattainable and therefore specious. No power on earth can protect the United States from terrorist attack, and during the ten days prior to Ashcroft's pledge of renewed vigilance, three senior administration officials found time in their busy schedules to inform the American television audience that such attacks were both imminent and inevitable. Vice President Dick Cheney on May 19: "The prospects of a future attack against the United States are almost certain . . . not a matter of if, but when." Robert Mueller III, the director of the FBI, on May 20: "There will be another terrorist attack. We will not be able to stop it." Donald Rumsfeld, secretary of defense, on May 21: "It is only a matter of time. . . ."

The reorganization of the FBI thus proceeds from a false first premise, and as compensation for both its past and future failure the Justice Department asks to be rewarded with broader authority, more money, and better flowcharts. The White House and the Pentagon meanwhile issue statements reaffirming the government's commitment to send flowers and attend funerals.

COMBAT PUBLIC CORRUPTION AT ALL LEVELS

Some levels, not all levels. When it comes to the arrest and detention of bearded Arabs carrying explosives and harboring deformed political ideas, the FBI can be counted upon to act with the vigor that resulted in the incineration of seventy-six Branch Davidians, most of them women and children, in a compound near Waco, Texas, in April 1993. I don't think that we can expect the Bureau to bring the same due diligence to the investigation of well-placed corporate executives who rob their shareholders, falsify their tax records, participate in the drug and weapons trades. The Enron Corporation rigs the price of electricity to steal billions from the public purse in California, but Enron's close association with the Bush Administration is of as little interest to the FBI as the administration's dealings with some of the same Saudi sheikhs who refocus their own resources on the "frontline positions" of Al Qaeda.

REFORM THE FEDERAL BUREAU OF INVESTIGATION TO PUT PREVENTION OF TERRORISM AT THE CENTER OF ITS LAW ENFORCEMENT AND NATIONAL SECURITY EFFORTS

The directive recurs, either verbatim or in close paraphrase, in all the texts distributed on May 30 (the Fact Sheets, Ashcroft's statement, the guidelines pertaining to "The Use of Confidential Informants," "Terrorism Enterprise Investigations," "General Crimes," and "Undercover Operations"), and in every instance it serves either as an excuse for incompetence or as a claim to extrajudicial privilege.

The attorney general didn't name any names or lay the blame at the feet of godless politicians, but neither did he try to hide the thought that if the FBI had been allowed to do its job the twin towers might still be standing watch over the lower reaches of the Hudson River. Unfortunately for America and the American way of life, the Bureau for the last thirty years has been crippled by "organizational and operational restrictions," bound up in "unnecessary procedural red tape." Never mind that the restrictions were imposed for good reason in the early 1970s when it was discovered that the FBI had been running illegal intelligence operations against as many as 2 million American citizens suspected of opposing the war in Vietnam or supporting the Civil Rights Movement.

What the Fact Sheet bills as the bold shifting of emphasis "from prosecution to prevention" unbinds Prometheus, removes the hobbles of bureaucratic restraint, and puts an end to the timid and overly literal-minded deference to the language of the Bill of Rights. No longer will the FBI's 11,000 agents sit feebly in their chairs filling out forms, asking permission to look out the window, waiting to "sift through the rubble following a terrorist attack"; they will "intervene early and investigate aggressively where information exists suggesting the possibility of terrorism."

I've never been much good at deciphering the small print in government documents, but unless I miss the point of the guidelines on "Terrorism Enterprise Investigations," they grant the FBI license to commit crimes when and if the circumstances warrant an especially strong defense of the public safety and the common good. The more serious felonies, of course, still require authorization from FBI headquarters in Washington or the Special Agents in Charge of a regional office, but such authorization is not to be unreasonably withheld (not when innocent American lives are at stake), and if time doesn't allow for a written confirmation the money laundering, the giving and taking of bribes, the winking at murder, the search, seizure, and entrapment of likely suspects can go forward on a word whispered into a secure telephone or an unmarked car.

PROTECT CIVIL RIGHTS

Among the new FBI special units put in place since last September 11, the Fact Sheet happily lists "the Financial Review Group, the Document Exploitation Group, and E-mail Exploitation Group . . . the Telephone Applications Group, as well as the Threat, Warning, Analysis and Dissemination Groups." Brisk, efficient, and blandly abstract, the terminology favors the classification of civil rights as nuisances that get in the way of law-enforcement officers rummaging through bank records and lingerie drawers in order to protect the American people from the swarm of terrorists in their midst. Ashcroft spoke to the need for a new understanding of the word "protection" by explaining, once again, that the onerous regulations under which the FBI has been operating for the last thirty years "mistakenly combined timeless objectives—the enforcement of the law and respect for civil rights and liberties—with outdated means."

As modified by the context and subject to the circumstances, the phrase "outdated means" can be taken to refer to any paragraph in every article of the Constitution. Which is both good to know and important to bear in mind because a modern war against terrorism cannot be fought with an old scrap of parchment and obsolete notions of liberty. Let too much liberty wander around without an escort and who knows when somebody will turn up with a bread knife or a bomb. Better to remember the lesson learned in the Vietnam War, which proved that the best way to save the village was to destroy it. So also now, in another time of trouble, the American people can best preserve their liberties by sending them to a taxidermist or donating them to a museum.

ENCOURAGE CITIZENS TO JOIN LAW ENFORCEMENT IN BEING VIGILANT AND WATCHFUL FOR SUSPICIOUS ACTIVITY

A casting call for informants of every known description—for neighborhood gossips and public scolds as well as for professional criminals and amateur conspiracy theorists. The Fact Sheet notes both a website and a toll-free telephone number, together with the heartening news that in the months of April and May the FBI received 225,000 tips by email, 180,000 tips over the phone. No power point indicates the number of agents assigned to the task of sorting out the false rumors from the groundless suspicions and the vengeful slanders; nor do the guidelines refer to the number of agents currently listening to wiretaps and peering at the film footage captured by the several million surveillance cameras now stationed at all points on the American compass. Constant supervision on so vast a scale must add to the burden of "unnecessary procedural red tape," which presents a contradiction again suggesting that Ashcroft's reorganization of the FBI serves a purpose other than the one announced at press conferences. A well-ordered police state rests on the cornerstone of a cowed citizenry, and how better to promote a decent respect for authority than by encouraging people to imagine themselves wearing a sheriff's badge, a well-tailored uniform, and a pair of polished boots.

As has been said, I don't rate myself an accomplished interpreter of government-sponsored prose, and my reading of last May's Justice Department text is possibly too naive. The country finds itself confronted by unscrupulous enemies who mean to do us no small harm, and I don't doubt that the FBI's revised guidelines will equip many of the Bureau's undoubtedly beleaguered agents with the improved means to perform an urgent and all but impossible task. But neither do I put much faith in Ashcroft's declarations of democratic principle or intent, and too many of the Bureau's reformulated purposes fit too neatly with the Bush Administration's wish to set itself above the law—to cloak its actions in the veil of sovereign secrecy and to accuse its political opponents of base or disloyal motives. Often when watching the administration's chief executives address a television audience I'm reminded of corporate lawyers talking to a quorum of recently bankrupted shareholders, and usually I'm left with the impression that they would like to put the entire country behind a one-way mirror that allows the government to spy on us but prevents us (for our own good, of course, and in the interest of national security) from seeing it.

What I find surprising is the lack of objection. The opinion polls show four of every five respondents saying that they gladly would give up as many of their civil rights and liberties as might be needed to pay the ransom for their illusory safety. The docile Congress takes its cue from the polls, and the news media rush to produce movies and television dramas (six of them running as weekly serials as of May) that portray the agencies of government (the CIA, the Supreme Court, the White House, and the United States Navy) as the institutional equivalents of Tom Cruise.

Given the amount of political capital that can be raised from an electorate sedated with the drug of fear, it's no wonder that our field commanders on the podiums in Washington never tire of discovering, once and sometimes twice a week, yet another terrorist threat lurking under the Brooklyn Bridge, disembarking from a plane in Chicago, driving a rented truck north toward Boston or south to Tallahassee. Nor is it any wonder that some of the leading figures in the administration don't bother to hide their disdain for an audience so easily deceived. When I listen to Attorney General Ashcroft, I'm never sure whether he intends an artful lie or believes himself to be relaying a message from God, but when I listen to Secretary Rumsfeld or Vice President Cheney, I know that I'm in the presence of cynical politicians who enjoy playing the game of Washington charades. Rumsfeld is particularly good at the tone of contemptuous irony. Last autumn, when the U.S. Air Force was bombing Afghanistan, Rumsfeld often appeared at the Pentagon press briefings to jokingly disparage the disinformation that he was passing off as truth. In Brussels for a meeting of the NATO allies in early June he elaborated the style of his performance, and to a crowd of reporters asking about the progress of the war on terrorism, he delivered a speech worthy of the riddling fool in one of Shakespeare's enchanted forests:

> The message is that there are no "knowns." There are things we know that we know. There are known unknowns. That is to say there are things that we now know we don't know. But there are also unknown unknowns. There are things we don't know we don't know. So when we do the best we can and we pull all this information together, and we then say well, that's

basically what we see as the situation, that is really only the known knowns and the known unknowns. And each year, we discover a few more of those unknown unknowns. . . . There's another way to phrase that and that is that the absence of evidence is not evidence of absence.

Just so, and no more questions need be asked about why you and I and little sister Susie will soon be going off to jail. "The absence of evidence is not evidence of absence," and although none of us knows that we've been thinking unknown Arab thoughts (Susie's only twelve), Mr. Rumsfeld knows, and so does Mr. Ashcroft; they've been listening to the tapes and looking at the pictures, and there among the power points, the absent evidence, and the known unknowns, they've found us in a tent with a camel and Scheherazade.

August 2002

Light in the Window

You may not be interested in war, but war is interested in you.
—Leon Trotsky

As a promotional venue for any season's collection of worthy thoughts and tasteful sentiments, the Sunday *New York Times Magazine* commands the authority of the show windows at Bergdorf Goodman. Of the moment and with the trend, the editors arrange the sociopolitical merchandise in ways meant to attract discriminating shoppers in the markets of received opinion—well-informed and right-thinking people, competently educated and decently affluent, alive to the similarities in the works of Versace and Matisse, fond of animals and the several shades of beige. Although the editors occasionally make space for ideas a trifle too advanced for some of their less sophisticated readers in Oklahoma or eastern Queens, they don't take chances with the big-ticket items or with what they judge to be the consensus of uptown money and downtown style.

Which is why, on first glancing at the no-nonsense cover lines for the issue of January 5—"The American Empire (Get Used to It)"—I knew that I was in the presence of an important fashion statement. The United States Army (very with it, very now) was on its way to an invasion of Iraq, there to exhibit a modish line of summer weapons at the military equivalent of a runway show, and the *Times* had gone to the trouble of furnishing a helpful program note: What to watch for, when to applaud, how to think about this year's new and exciting look in geopolitics. Michael Ignatieff, a brand-name foreign-policy intellectual recruited from the faculty of Harvard University, matched the assertive tone of his lead article to the red, white, and blue block lettering (not fussy, very bold) of the magazine's cover art:

> Americans are required, even when they are unwilling to do so, to include Europeans in the governance of their evolving imperial project. The Americans essentially dictate Europe's place in this new grand design. The United States is multilateral when it wants to be, unilateral when it must be; and it enforces a new division of labor in which America does the fighting, the French, British and Germans do the police patrols in the border zones and the Dutch, Swiss and Scandinavians provide the humanitarian aid.

An editor's note identified Ignatieff as the director of the Carr Center at the Kennedy School of Government, also as a teacher of "human rights" well versed in the syllabus of the world's sorrow. A man of sense and sensibility who had spent a lot of time "walking around" in the "frontier zones of the new American empire" in Bosnia, Kosovo, and Afghanistan, and who knew, as he himself said, that it wasn't enough to draw "the big picture" on the wall of a classroom in Cambridge, that one really must "get out of Harvard Yard" if one wants to "get really close to the intimate, tragic detail of it all."

And what did he learn, the professor, from his poking around in Afghan tents and Balkan graves? If nothing else, how to write sententious and vacant prose, most of it indistinguishable from the ad copy for an Armani scarf or a Ferragamo shoe. Too much direct quotation from the professor's text might be mistaken for unkindness, and I enter three of his obiter dicta into the record only because they fairly represent the attitudes currently in vogue among the marketers of the country's preferred wisdom:

> Imperial powers do not have the luxury of timidity, for timidity is not prudence; it is a confession of weakness.
>
> [The United States] remain[s] a nation in which flag, sacrifice and martial honor are central to national identity.
>
> The question, then, is not whether America is too powerful but whether it is powerful enough. Does it have what it takes to be grandmaster of what Colin Powell has called the chessboard of the world's most inflammable region?

If Ignatieff doesn't for a moment doubt that America has what it takes to play chess with Genghis Khan or Darius the Great, neither does he say anything that hasn't been said, repeatedly over the last nine months, by the cadre of Washington propagandists, both Democrat and Republican, writing for the policy journals that supply the government with its think-tank thoughts and Sunday-morning media phrases—America, the world's unrivaled hegemon, an empire in fact if not in name, its sovereign power the only hope for less fortunate nations groping toward the light of free markets and liberal democracy. Be not timid, do not flinch. Shoulder the burden of civilization and its discontents. Lift from the continents of Africa and Asia the weight of despotic evildoers. Know that if America does the fighting, other people will do the dying. Learn to appreciate the refined elegance (conceptually minimalist, gracefully postmodern) of high-altitude precision bombing.

When delivered by one of the rabid polemicists allied with Lockheed Martin, Fox News, or the Baptist church, the same message usually comes with a tactical objective in view—the seizure of the Iraqi oil fields, the destruction of Hugo Chávez, the safety of Israel. Ignatieff imparts to it an air of languid abstraction, not wanting to disturb anybody with the "intimate, tragic detail of it all." He briefly raises the question of terminology (America as democratic republic, America as military empire), but then he goes on to say that it doesn't make much difference how America chooses to see itself. The words don't matter. The country is what it is, so rich and powerful and good that it can't help but do what is just and right and true.

Never having met Ignatieff or read his books, I don't know how he defines his politics, or whether he construes himself as a liberal, a conservative, a lapsed Marxist, or a reconstructed Tory. After making my way through the 7,000 words of his article for the *Times,* I still couldn't guess who were his enemies and who were his friends, and it occurred to me that much the same can be said about most of the government officials and high-end journalists whose commentaries have decorated the display windows of the national news media for the better part of the last twenty years. They strike poses and adopt attitudes, pleased to imagine that politics amount to little else except the staging of *tableaux vivants,* the crowd scenes and the musical accompaniment matched with the power points in the season's polling data. The reliance on theatrical effects long ago destroyed the credibility of the voices of conscience associated with liberal causes and the Democratic Party. Too many nominally left-wing defenders of the realm couldn't disguise their loyalty to the right-wing Mouton Rothschild, their rhetorical tours de force (on behalf of racial equality, social justice, freedom of expression, etc.) too laughably at odds with their views of the sea from a sundeck in

East Hampton. A similar fate befell the apostles of the conservative truth soon after the election of Ronald Reagan. The actor arrived in Washington under the impression that he had been hired to make a movie, and it was only a matter of months before the claque of his apologists in the news media began to think of aircraft carriers as set decorations.

Reading Ignatieff I was reminded of a dinner-table conversation in Washington in the middle 1980s at which an authoritative syndicated columnist explained that he was "depressed" by "the quality of the regime" in Nicaragua. Judging only by the tone of his voice, I might have guessed that he was talking about a second-rate wine or a Caribbean resort hotel gone to seed and no longer fit to welcome golf tournaments. He wasn't concerned about Nicaragua's capacity to harm the United States; the army was small and ill equipped, the mineral assets not worth the cost of a first-class embassy. Nor did the columnist think the governing junta particularly adept at exploiting "the virus of Marxist revolution." What troubled him was the "indecorousness of the regime." Nicaragua was in bad taste.

In the similarly detached context of a Fifth Avenue cocktail party on the Tuesday after the publication of Ignatieff's article, I listened to two political correspondents, one from *Vanity Fair* and the other from *Newsweek,* analyze the differences between the Democratic and Republican attitudes toward the liberation of Iraq. The important question was aesthetic, not geopolitical; not whether Saddam Hussein deserved to die a coward's death (the fact so obvious that it didn't bear discussion) but how to accessorize the coverage of the assault on Baghdad, whether it was best to present it as a crusade ordered by God or as a hostile corporate takeover arranged by a consortium of Texas oil companies. The first option indicated a heartland sensibility, the second an instinct perversely urban.

I mention the conversation because, like the brass-band promotion of Ignatieff's article in *The New York Times Magazine,* it speaks to the insouciance with which so many people party to the formulation of our public argument regard the current recasting of the country's laws. The government in Washington makes no secret of its wish to eliminate the freedoms of speech, thought, and movement synonymous with the workings of a democratic republic, and where is the senator or syndicated columnist who gives voice to an intelligible objection? To the civilian populations of Basra or Islamabad it might not make much difference whether the United States styles itself hegemon or republic—the bombs fall from the same clouds, the buildings collapse into the same ruins, and who doesn't know that an F-16 by any other name is still an F-16? For the residents of Philadelphia and Colorado Springs, the words should matter; despite Ignatieff's bland assurances to the contrary, so should their supporting connotations. A government that becomes accustomed to thinking of itself as an empire falls easily into the habit of issuing imperial decrees and soon acquires the characteristics that Secretary of State Colin Powell last February attributed to a failed state, "unrepresentative of its people . . . rife with corruption," blighted by "a lack of transparency," thinking that "it can achieve a position on the world stage through development of weapons of mass destruction that will turn out to be fool's gold . . ." The secretary was speaking of North Korea and Iraq; he might as well have been talking about Vice President Dick Cheney's vision of a reconfigured United States.

If not as a concerted effort to restrict the liberties of the American people, how else does one describe the Republican agenda now in motion in the nation's capital? Backed by the specious promise of imminent economic recovery and secured by the guarantee of never-ending war, the legislative measures mobilized by the White House and the Congress suggest that what the Bush Administration has in mind is not the defense of the American citizenry against a foreign enemy but the protection of the American oligarchy from the American democracy.

As if wishing to leave nobody in doubt about the political bias now afoot in Washington, President Bush took the trouble to juxtapose his endorsement of affirmative action for the rich (the speech to the Economic Club of Chicago on January 7 favoring the removal of all taxes paid on corporate dividends) with his objection to affirmative action for the poor (his remarks from the White House on January 15 finding fault with the admissions policies at the University of Michigan). The interval of a week between the two announcements was brief enough to impress the lesson upon a national television audience known for its short attention span, but among most of the upscale journalists in New York (of the moment, with the trend) the point was by and large ignored. The best and most tasteful opinion doesn't countenance the notion of class warfare. President Bush rejects even the suggestion of such a thing as wrongheaded and maybe treasonous, partisan agitprop distributed by envious Democrats and would-be demagogues. His indignant tabling of the proposition, in a speech on January 9 at a Virginia company that produces American flags, was strongly seconded by the right-thinking managers of the nation's better media boutiques, who regard the subject as preposterously *démodé*—threadbare cant found in the attic of the 1960s with the rest of the sensibility (go-go boots, *Sgt. Pepper,* Woodstock, Vietnam) that embodied the failed hopes of a discredited decade.

The media take pride in their exquisite collections of historical certainty, and so, being persuaded that class war invariably manifests itself as an uprising of the angry poor against the greedy rich (pitchforks, sansculottes, the guillotine), and having seen the reassuring photographs of both Teddy and Franklin Roosevelt, they also know that all American politicians are, by definition, gregarious and open-hearted people, industrious, well meaning, occasionally eccentric but always friendly, sometimes caught, unwillingly and through no fault of their own, in the webs of corrupt circumstance. Armed with combat-hardened anecdotes excerpted from the writings of Tom Clancy and the late Stephen Ambrose, our high-toned window dressers like to imagine that politics are about the more or less attractive arrangement of words, not about who gets to do what to whom, at what price, and for how long. Their gift for clever decoration serves the interest of the oligarchy currently at home in Washington (more frightened of the freedom of the American people than of the tyranny of Saddam Hussein), and obscures the fact of a war waged by the angry rich against what they perceive to be the legions of the greedy poor.

March 2003

Yankee Doodle Dandy

McKinley understood quite well that Americans might accept, albeit uneasily, an accidental empire; an empire by design they would not have borne, not even in the giddy, war-feverish days of 1898. Like the buncombe artist who cranked the handle that operated the "Wizard" of Oz, so McKinley now cranked the handle of "destiny," set in motion the "march of events," and manipulated the "hand" of the "Almighty," which was no more than an empty glove.
—Walter Karp

*D*uring the eight months prior to the invasion of Iraq, the American news media were content to believe the government's fairy tale about its reasons for sending the tanks eastward into Eden. The Bush Administration's buncombe artists could tell any story they pleased about Western civilization being held for ransom by Saddam Hussein's weapons of mass destruction, and even when the plotlines were shown to depend upon suborned testimony and counterfeit intelligence, the media vouched for the wisdom of Oz. Why not? What was to be gained by casting doubts? The fairy tale sold newspapers, boosted television ratings, curried favor at the White House and the FCC, drummed up invitations from the Pentagon to attend the military costume party in the Persian Gulf.

Three months after the capture of Baghdad the fabulous weapons were still nowhere to be found, and the government ventriloquists in Washington—not as fortunate as President William McKinley in their choice of a splendid and self-serving little war—were experiencing technical difficulties with the empty glove. Unable to find in it anything that resembled the hand of God, they classified impertinent questions as proofs of disloyalty or ingratitude—not fair to the troops (brave Americans one and all), insensitive to the plight of the Iraqi people (formerly enslaved, now free to elect any imam who promised not to slit their throats), the arsenal of doom never the only reason for the advance into the valley of the Euphrates (merely one of many reasons, the others all very geopolitical and complex but failing to meet the necessary quotas of fear and loathing), the world well rid of Saddam Hussein no matter what the pretext for his departure. In the near term and at least for the time being, the excuses were sufficient to their purpose. A few Democratic politicians demanded extensive public hearings, but, given the Republican majorities in Congress, the prospect was unlikely. Newt Gingrich, former Speaker of the House and lately a prominent front man for the product of American empire, found comfort in an opinion poll showing 60 percent of the American electorate accepting the lie about Iraq's alliance with Al Qaeda, and therefore an accessory to the crime of 9/11. "On this one," he said, "the president is 99% safe."

Nor were the news media inclined to upgrade the president's cynicism and dishonesty into a story worthy of a ranking with the seduction of Monica Lewinsky. *Time* magazine devoted a worried headline to the "Weapons of Mass Disappearance," and a scattering of critics in the literary press characterized preemptive blitzkrieg as a betrayal of America's best interests and dearest principles; in the journals of large circulation only Paul Krugman in the *New York Times* raised a similar point in language unsweetened with apology. Taking note of our government's gift for "systematically and brazenly" distorting the facts "to an extent never before seen in U.S. history," Krugman went on to say, "suppose that this administration did con us into war. And suppose that it is not held accountable for its deceptions. . . . In that case, our political system has become utterly, and perhaps irrevocably, corrupted."

I don't argue with Krugman's judgment of the dissembling mountebanks, among them our secretaries of state and defense, busy cranking the handles of "destiny" and setting in motion "the march of events," but I'm reluctant to concede the phrases "perhaps irrevocably" and "never before seen in U.S. history." Krugman's suspicions are well placed but a hundred years behind the news. The government in Washington seldom lacks for a quorum of cheats and liars; the con games take similar and traditional forms, and as an appreciation of the one currently in progress, I know of none better than the late Walter Karp's *Politics of War*.

First published in 1979, the book describes the emergence of the United States as a world power between the years 1890 and 1920— our contrivance of the Spanish-American War and our gratuitous entrance into World War I—and by filling in the back story of an era in which "mendacious oligarchy" organized the country's politics in a manner convenient to its own indolence and greed, Karp offers a clearer understanding of our current political circumstance than can be found in any two or twenty of the volumes published over the last ten years by the herd of Washington journalists (milk fed, free of hoof-and-mouth disease) grazing on the White House lawn. The leading characters in Karp's narrative belong to another century, dressed in costumes no longer sold in stores, but because he was both a gifted writer and a careful historian the story is alive and well and still present on the page. His passion was politics, and his precepts were simple and few. He believed that in America it is the people who have rights, not the government, and he made a clear distinction between the American republic and the American nation—"deadly rivals for the love and loyalty of the American people." Always mistrustful of what he called the "official version" of things, Karp professed his allegiance to the republic, which he understood as a body of law fitted to a common interest and a human scale, shaped by the spirit of liberty in which the country was conceived. The nation he regarded as a poor dim thing, assembled as a corporate entity, sustained by an "artificial patriotism," and given the semblance of meaning only when puffed up with the parade music of a foreign war.

In the inevitable and implacable conflict between the two ideas of America, Karp saw the "great drama" of the country's political history, and the sequence of events accounted for in *The Politics of War* provides him with a demonstration of his thesis and an occasion for both his eloquence and his sardonic wit. He revises the conventional portraits of Wilson as a principled idealist, of McKinley as a hapless incompetent, of American policy as a message of Christian goodwill, of the American press as the champion of truth, and to read his deconstruction of Wilson's glib and pious oratory—the "dictates of humanity," "peace without victory," etc.—is to hear the similarly unctuous sound in President George W. Bush's declarations of war against all the world's "evildoers" in the name of God and "all mankind." Portraying the character of the lying Washington pantaloons, aligned in the 1890s with the Republican money power, among them Nelson Aldrich of Rhode Island ("who frankly viewed any kind of politics save the politics of corrupt privilege as sentimental rot"), Karp might as well be sketching the bandmasters of the Bush Administration, who make no secret of their contempt for the American republic and mint the hard coin of the public trust into the debased currency of their private ambition.

The Politics of War brings to bear the clarity of hindsight on the chicanery of the present, and by so doing answers questions never asked by the *Wall Street Journal* or dreamed of in the philosophy of CNN. Just as Operation Iraqi Freedom was not about the rescue of the Iraqi people, so also the Spanish-American War was not about "the sacred cause of Cuban independence," and our entry into World War I not about making the world "safe for democracy." Presidents Wilson and McKinley sought to punish foreign crimes against humanity (the ones committed by villains in Brussels and Havana) in order to make America safe for the domestic crimes against humanity committed by fine, upstanding, corporate gentlemen in Boston and Chicago. If by 1890 the Industrial Revolution had made America rich, so also it had alerted the electorate to the unequal division of the spoils. People had begun to notice the loaded dice in the hand of the railroad and banking monopolies, the tax burden shifted from capital to labor. A severe depression in the winter of 1893–94 brought with it widespread unemployment, vicious strikes in the Pennsylvania steel mills and West Virginia coal mines, hobo armies on the march in the Ohio Valley and the Appalachian Mountains. The demand for social and political reform prompted the angry stirring of a Populist movement across the prairies of the Middle West, and as a cure for the distemper of an aroused citizenry—"something," in the words of an alarmed U.S. senator, to knock the "pus" out of this "anarchistic, socialistic and populistic boil"—the McKinley Administration came up with war in Cuba, the conquest of the Philippines, the annexation of Puerto Rico, and an imperialist foreign policy deemed "essential to the greatness of every splendid people," necessary "to the strength and dignity of any nation." Only by infecting the republic with the delusion of imperial grandeur could the nation (which exists only in relation to other nations) smother the republican spirit and replace the love of liberty with the love of the flag—every true American a patriot, all political quarrels to be suspended in the interest of "the national security."

By the time that President Theodore Roosevelt moved his cavalry horses into the White House stables in 1901, the last remnants of populist unrest had drifted into the sunset with the wreckage of the Spanish fleet, and for the next five years the agents and apostles of the American nation gloried in a triumph of wealth and cynicism presumed sufficient to silence any loose-mouthed or ill-bred talk about ordinary citizens deserving a say in a government nominally democratic. The presumption soon collapsed under the weight of its complaisance and stupidity. Incapable of managing an economy that it could only prey upon, "the money power" and its hired politicians consigned the arrangement of the country's financial affairs to a consortium of swindling bankers and bribed legislatures, and by 1906 the continuing proofs of the oligarchy's disdain for such a thing as the common good translated the resentments once lodged in the rural counties of Populist discontent into the muckraking politics of the Progressive movement. Against an urban and more sophisticated opposition, the friends of the protective tariff and Yankee Doodle Dandy had need of more strenuous countermeasures, and Karp devotes the greater part of his book to the "bottomless" deceit of Woodrow Wilson, who engineered America's entry into World War I in order that he might play the part of a great statesman settling upon the tumult of nations a peace deserving of comparison to an act of God. Without an army in Europe the president couldn't strike as handsome a pose as the one arranged for President Bush on the deck of the U.S.S. *Abraham Lincoln,* and if the setting up of the photo op required the deaths of 100,000 American soldiers in the mud of France, the sacrifice could be written off to America's privilege "to spend her blood and her might for the principles that gave her birth." The falsity of Wilson's pose as a man of elevated principle and noble character moves Karp to savage mockery, and a single sentence can be taken as indicative of both the sense of his argument and the strength of his prose:

> The decisive trait of Wilson's political character was vainglory: a hunger for glory so exclusively self-regarding, so indifferent

to the concerns of others, that it would lead him to betray in turn the national movement for reform, the great body of the American people, the fundamental liberties of the American Republic, and in the end the hopes of a war-torn world.

On almost every page of *The Politics of War* an attentive reader can find the lines of connection and resemblance between time present and time past—the fortuitous sinking of the *Maine* in February 1898 and of the *Lusitania* in May 1915 providing the promoters of American empire with the same sort of casus belli as was presented to the Bush Administration by the destruction of the World Trade towers in September 2001; Spain's fifth-rate colonial power in Cuba depicted by the McKinley Administration as "the most wicked despotism there is today on this earth"; both Wilson and McKinley making campaign promises equivalent to Bush's compassionate conservatism and just as promptly rescinding them once they had been elected to office; Teddy Roosevelt at Fourth of July picnics denouncing "ultra-pacifists," "poltroons," and "mollycoddles" in the manner of Rush Limbaugh excoriating Susan Sontag and Harvard University; the many similarities in the distribution of alarmist propaganda— Germany in the summer of 1915 said to be capable of landing, in a matter of sixteen days, 387,000 troops on the coast of New Jersey, Saddam Hussein's weapons of mass destruction in the autumn of 2002 said by British intelligence to be ready for use within a matter of forty-five minutes; both the McKinley and the Wilson administrations served by a warmongering press ascribing atrocities to Spanish viceroys (Cuban peasants fed to sharks) and to German generals (Belgian nuns roasted over burning coals) as promptly on cue as Fox News charging Saddam Hussein with the burial of Iraqi infants in mass graves; the Congress in June of 1917, in an atmosphere as clouded by militant paranoia as the mind of Attorney General John Ashcroft, passing an Espionage

Act under which any and all criticism of the government acquired the distinction of a felony.

Walter Karp died in 1989, at the age of fifty-five of wounds inflicted by clumsy doctors in a careless hospital, but had he lived long enough to marvel at the wonderful wisdom of Oz now being slopped into the pails on the White House lawn, I expect that he would have welcomed the chance to embellish the likenesses among three American presidents, each of them seen in the warm glow of their radiant hypocrisy, doing their eager and patriotic best to place the country on the marble footings of self-righteous oligarchy. He was a writer who could count among his antecedents dissenting spirits as passionate and as troublesome as those of Ambrose Bierce and Mark Twain, and if somewhere not too far offstage I can still hear his antic improvisations on the theme of "elective despotism," I also remember that he was fond of citing Thomas Jefferson's dictum that "we are never permitted to despair of the commonwealth." Karp saw the American nation as a sickness in the body of the American republic and therefore subject to cure; he pursued the study of history as a means of protecting the future from the past, and by way of shoring up his hope for better days to come he relied on the memory of worse days providentially gone.

August 2003

Wild West Show

It is not easy in any given case—indeed it is at times impossible until the courts have spoken—to say whether it is an instance of praiseworthy salesmanship or a penitentiary offense.
—Thorstein Veblen

When President George W. Bush moved into the White House in the winter of 2001, he let it be known that he intended to run the government as if it were a business, and two years later I don't know why it comes as a surprise that the ten-year federal budget projection has been reduced from a $5.6 trillion surplus to a $4 trillion deficit, or that our splendid little war in Iraq turns out to have been sold to the American public in the manner of a well-promoted but fraudulent stock offering. The man has been true to his word, the corporation of which he deems himself chairman and chief executive officer not unlike the ones formerly owned and operated by his friends, fund-raisers, and fellow bandits at Enron and Arthur Andersen. The administration's economic and military schemes rely for their success on budget analysts who reconfigure debt as credit, on auditors at the intelligence agencies who rig their balance sheets with sham transactions (for African uranium), false data (establishing Saddam Hussein's connection to Al Qaeda), offshore special-purpose entities (to contain the otherwise invisible weapons of mass destruction).

The acknowledgment in early July of the president's misstatement about the African uranium prompted the members of Washington's political theater company to strike strong poses of indignation and dismay, to call for congressional investigations, and to bat the tennis balls of blame around the Sunday talk-show circuit. Condoleezza Rice, the national security adviser, told the cameras at Fox News that the suspect information had come from the British government; on NBC's *Meet the Press* as well as on ABC's *This Week,* Secretary of Defense Donald Rumsfeld said that "it's not known" what is or was "inaccurate," and to watch the ministers of state search for ways around the truth was to be reminded of the swarm of thieving corporate executives hustled into courtrooms over the last two years in the hope that they might recall what happened to Global Crossing's once-upon-a-time $47.6 billion of market value or to the 59,000 employees who used to work for Kmart. Nine times out of ten they met the questions with answers matched in quality to Vice President Dick Cheney's best guess as to the whereabouts of Osama bin Laden.

Because the Bush Administration's modus operandi resembles that of a corrupt monopoly (publicly owned but privately managed), much of its domestic and foreign policy can be understood in terms of the hidden surcharge and the dishonest annual report. Fraud is another word for freedom, "to make a killing" the highest form of patriotism or praise. Mark up the price of the American military occupation of Iraq from $2 billion to $4 billion a month, or guide WorldCom into the desert of a $9 billion accounting error, and whether it is Donald Rumsfeld or Bernard Ebbers explaining the arithmetic to Tim Russert or a judge, the story follows a by now familiar script—the once trusted brand-name corporation becomes entangled in the nets of perjury and graft, assets worth $100 million depart for points unknown, the stock price falls from $95 to 30 cents a share, the company's pensioners cast adrift in open boats with a cup of rainwater and a scrap of raw fish. The high-end executives meanwhile cash their stock options at the best possible price, reward their own grand and petty larcenies with severance payments in the amount of $30 million or $40 million, retain the apartment in Paris and the bank account in Zurich, go off to Colorado with the golf clubs and the skis. Transfer the procedure from the private to the public sector and the government fattens the defense budget by sending the soldiers and the tanks out to pasture in Iraq, clears the Pentagon's inventory of weapons no longer stylish, and distributes the reconstruction contracts to Bechtel, Halliburton, and any other friend of liberty willing to lend a hand with the oil derricks around Baghdad and the balloons at next year's Republican nominating convention.

The administration's current tax law could have been written (possibly *was* written) by a caucus of fund managers at Merrill Lynch or Goldman Sachs, less a matter of poor policy than of outright theft and solidly grounded on the time-honored principle beloved by generations of Wall Street stock touts—steal from the poor to feed the rich, and if asked to explain the mechanics of the deal, talk about family values. When the president proposed the legislation last February, he said that 92 million Americans would receive an average tax reduction of $1,083. A fine sentiment that Veblen undoubtedly would have recognized as an instance of "praiseworthy salesmanship," but, like the statement about the African uranium, not true and probably better understood as a "penitentiary offense." For families with incomes of between $30,000 and $40,000 the dividend reduction amounts to $24; in the safer parts of town where the wine comes in a bottle with a cork, John Snow, the secretary of the treasury, receives a refund of $275,000; Defense Secretary Rumsfeld, the sum of $184,000. The high-end campaign contributors meanwhile shop for another beach house in sunny Florida, and President Bush tours the country with a message from the Bible, poses for photographs in front of signs bearing the motto CORPORATE RESPONSIBILITY, saying to the newly out-of-work clerks and factory hands assembled in the convention center, "I believe people have taken a step back and asked, 'What's important in life?' You know, the bottom line, and this corporate America stuff, is that important? Or is serving your neighbor, loving your neighbor like you'd like to be loved yourself?"

Given the transparency of the Bush Administration's lies, the unctuousness of its voice, and its unilateral contempt for any rule of law—civil, international, moral, or financial—why is it that we don't make unto the Lord a not-so-joyful noise? It's neither accident nor coincidence that as the stock market over the last two and a half years has lost roughly $6 trillion in value, so also the unemployment rate now stands at its highest level in a decade (6.5 percent), and the current budget deficit of $450 billion at the highest level in the country's history. We see before us a government fitted to the specifications of not just any corporation (most of which preserve some sense of obligation to the public welfare and the common good) but an especially rapacious corporation more nearly resembling a criminal syndicate, the risk remanded to the uninformed taxpayer instead of to the unwitting investor. Why then the audible silence in the news media and on the part of our nominally democratic politicians, and how does it happen that no crowd armed with lead pipes gathers in the union halls or the streets?

The customary answer comes in the form of a sermon about our all-American faith in a better future and a brighter tomorrow. The servants of the status quo shift the venue of possible complaint from a criminal to a civil jurisdiction, quieting the murmurs of objection with a flood of reassuring cant about the country's lack of class consciousness. The American oligarchy doesn't prey upon the American democracy because in America we have no such thing as an oligarchy. Perish the thought. What we have in this great country is what the Republican propagandists like to call "the Wild West of economies," the swinging door of a saloon on the old Arizona frontier, where sudden fortune might come for anybody with a $5 poker chip and a mule. The words to the wise invariably repeat the lesson taught by Norman Rockwell, Walt Disney, and the editors of the *Wall Street Journal*. A few liberties having been taken with the paraphrase, the standard text reads as follows:

> *Being American and therefore blessed at birth with the gene of egalitarianism, we don't envy people richer than ourselves; nor do we labor in the shadow of the false class distinctions that still cloud the mind of Old Europe. If a CEO earns 500 times as much money as one of his secretaries or assembly-line workers ($15 million as opposed to $30,000 per annum), it doesn't mean that the CEO is a selfish thug. The suspicion is invidious and French. The nice man owes his good fortune to hard work and his belief in God. We don't begrudge him his opulent lifestyle (too snooty, not enough time for his neighbors or his pet toad) because we know that soon, probably sooner than anyone thinks, we ourselves will become rich. In the meantime, while waiting for the report from the assay office or the coroner, we are expectant capitalists, devoid of resentment, happy to be living in a land of great abundance and unlimited opportunity. We are Americans, bless our hearts, just folks who would rather be shopping at Wal-Mart than at one of those fancy stores on Madison Avenue or Rodeo Drive, and anybody who wants to talk to us about class conflict or Karl Marx better damned well know that we don't hold the notion of a society that looks like some sort of English layer cake.*

I don't question the American generosity of spirit and natural gift of forbearance, but I think the Norman Rockwell portrait two or three generations behind the times, and I suspect that President Bush owes his standing in the opinion polls to our equally American inclination to make the criminal and the outlaw figures of romance. Whether cast as the hero or the villain of the tale, the man at ease with violence bends the rules to fit the circumstance, certain that his always noble ends justify his sometimes less than noble means. If we know the game is rigged, we also like to think that the owner of the roulette table loves his mother and looks like Michael Corleone.

President Bush walks the walk that can be imagined as once familiar on the old Chisholm or Santa Fe trail, entitled to his cocksure swagger by virtue of his having stolen his election to the presidency. The robbery stands on a par with Enron's looting of $28 billion from the California state treasury and admits President Bush to the long line of America's criminal ancestry that descends, with mounting degrees of firepower and computer capacity, from James Fenimore Cooper's Deerslayer to the Rocky Mountain fur traders, John Jacob Astor and the Sundance Kid, the Texas cattle barons and New York railroad kings, John D. Rockefeller, Boss Tweed, and Al Capone, Humphrey Bogart, Huey Long, Lyndon Johnson, Clint Eastwood, Michael Milken, Oliver North, Ivan Boesky, Tony Soprano, and others, as they say at the Academy Awards ceremonies, too numerous to mention.

Where else does Hollywood and the tabloid press find our company of heroes if not among the archetypal men on horseback who wander into dusty wooden towns to burn down the sheriff's office or eliminate the grinning Mexicans who have been terrorizing the dance-hall girls with their aimless and maniacal gunplay? When Frank and Jesse held up the Kansas City Fair in 1872, the local newspaper described the exploit as being "so diabolically daring and so utterly in contempt of fear that we are bound to admire it and revere its perpetrators," comparing the gang to the knights of King Arthur's Round Table. The national news media in 1960 borrowed the same romantic legend to herald the Kennedy gang's capture of a government suddenly bright with the banners of Camelot. The editors of *Fortune* and *Business Week* greeted the early success of Bernard Ebbers at WorldCom and Kenneth Lay at Enron with a similar fluttering of mindless awe—fearless CEOs, freebooting captains of commercial enterprise, tough-minded and resolute, driving up the stock price as if punching cows across the Missouri River into Marlboro country and the California energy market.

When too many proofs of government theft or corporate greed accumulate on the ledgers of the news (another defense contractor indicted for fraud, five more chief financial officers found to be in possession of indecent loans), the country's loyal media outlets stage the equivalent of a prairie revival meeting meant to call down the great spirit of financial reform upon the temples of Mammon. A choir of Harvard economists bears witness to the restoration of investor confidence; Alan Greenspan lowers the interest rate; four committees of Congress set forth to recover America's lost moral compass; a few sheepish perpetrators appear in handcuffs on CNN, Martha Stewart tied like a sacrificial goat to the post of a courthouse press conference; Sandy Weill, chairman of Citigroup, forbidden to talk to his own research analysts unless in the presence of a lawyer. George Tenet, the CIA director, takes a well-publicized fall for the mistake with the African uranium, and when the New York attorney general's office levies a $1.4 billion fine on the country's largest securities firms (among them Salomon Smith Barney and Morgan Stanley), the newspapers announce "the dawn of a new day on Wall Street."

The new day lasts for maybe a week or a month. Nightfall descends as abruptly on the New York financial markets as it does on the Brazilian rain forest, and although in our public discourse we like to bemoan the decay of conscience, in our private conversations, variations on the answer, "Yes, but I did it for the money," satisfy all but the most tiresome objections. Were Al Capone still alive he probably could count on steady work as a talk-show host on Fox News—an elder statesman, garrulous and wise, remembering the good old days in Chicago, offering his opinion as to whether money is better starched and laundered in the Cayman or the Channel islands.

President Clinton made it his practice to pardon friends convicted of felonies; President Bush makes it his practice to appoint them to federal office, and in California Arnold Schwarzenegger presents himself as a candidate for governor. To the extent that the romance of crime has become more popular over the last thirty years, in part the result of the continuous panorama of violence running around the glass walls of our news and entertainment media, we make heroes of the politicians and business executives who resemble the bounty hunters wandering in the wilderness of the old American West, predatory and nomadic figures holding to the rule "get in, get rich, get out," bored by the tasks of government, indifferent to the pleasures of civilization. If occasionally called upon to bring justice to Gold Hill or Silver City, they do so with a terrible and godlike vengeance. After the requisite number of killings, they depart (into the sunset, back to the aircraft carriers in the Persian Gulf), leaving to merely mortal men the tedious and un-American chores of settlement and burial. Although I don't doubt that many millions of voters admire President Bush for the handsome platitudes he distributes under the labels of "moral clarity," I suspect that an equal number of citizens place their trust in his work and record as an outlaw—maybe not up to the mark of Meyer Lansky or Shane but a gunsel reliable enough to ride with the Clanton gang and fit to hold the horses when the time comes to burn out the sheep herders or rob the train.

September 2003

The Golden Horde

*. . . the most difficult {decision} I've made in my entire life, except the one
I made in 1978 when I decided to get a bikini wax.*
—Arnold Schwarzenegger

*T*he well-contrived one-line joke was a little too well-rehearsed
and the winning smile adjusted from maybe one too many
angles in the green-room mirror—what a lovable guy, so wonderful
an action hero that he can play at being feminine and cute—but it
drew the preprogrammed laugh from the studio audience at the
taping of *The Tonight Show with Jay Leno* on August 6, and it served
as the announcement of an actor's wish to become governor of California. On cue and in time for the hourly headlines, the national
news media went for the story like an aquarium seal flopping after
the handheld fish. The tabloid press made loud and happy barking
noises—California's recall election compared to a Hollywood musical with a chorus of 135 dancing clowns; the Terminator loose
among the aliens in Benedict Canyon. The upscale parlor press
made mournful, elegiac sounds signifying the death of Thomas Jefferson and an end to "politics as we know it."

My own acquaintance in the profession inclines toward the latter company of scriveners (guardians at the gate of reason, protectors of the nation's moral fabric), which was why, during the eight days between Schwarzenegger's declaration of intent and the power failure that discomforted 50 million people across eight states on August 14 (an even more vivid reminder of the grim tidings likely to be hidden behind the curtain of tomorrow's news), it wasn't easy to avoid discussions about the flight of meaning from the American public square. No conversation was safe from earnest monologues accounting for the evils that had befallen the theory and practice of democratic self-government. Professors of political science professed themselves appalled. An alarmed publisher compared Schwarzenegger's effrontery to that of the emperor Caligula, who appointed his horse to a seat in the Roman senate. Various editors of the *Wall Street Journal* wondered what outrageous act of impudence our debased society might still recognize as a display of shameless gall. Nobody could think of anything other than cannibalism; lesser offenses against the sense of public decency merely would attract the customary book and movie deals. A wit suggested "Shameless Gall" as the name for a men's cologne. Noting that California wallowed in the quagmire of mindless ego, three contributors to the *New York Times* asked where were the "ropes of leadership" with which to drag the animal back to the high and solid ground of the "shining city on the hill."

The elevated tone of the conversations opened the question as to whether those present had spent the last twenty years somewhere outside the continental United States, or possibly in a library at Harvard with the collected works of James Madison. I couldn't quarrel with their definitions of effrontery or their observation that gravitas had escaped the clutches of Barbara Walters, but in what country did they think they were living, and how had they neglected to notice that the arguments put forward in *The Federalist Papers* had been replaced by the preening sophisms of Rush Limbaugh and

Howard Stern? The political ship of state designed to the specifications of the eighteenth-century Enlightenment sailed over the horizon in 1980 when Ronald Reagan was elected to the White House on the strength of his heartwarming sentiment and his talent for striking an heroic pose. How different Schwarzenegger's shameless gall from the shameless gall of President George W. Bush, who offered as his credentials for admission to the White House little else except a famous name, a record of failure and probable fraud as both a business executive and as a soldier, a friendly state of wholesome ignorance, and a desire for applause?

When temporarily at a loss for quotations from Cicero or Edmund Burke, the voices of conscience in New York blamed the country's troubles on the state of California. What were those people thinking out there in the golden West? Why were they so feckless, and where did they come by the idea that in the years between the Clinton and Bush administrations, Martin Sheen was the president of the United States? Because several of the people sitting at the table or the bar remembered that I happened to have been born and raised in San Francisco, their questions were often indistinguishable from accusations, and I soon found myself fortifying the specific answers with the bulwark of a general theory. The tendency is consistent with the California temperament and turn of mind.

Like Athena springing full-blown from the head of Zeus, California emerged full-blown from the myth of Golconda, its origin coincident with the discovery of gold at Sutter's Mill in 1848. The deposits of the miraculous, fortune-bearing sand ranged across an escarpment roughly 300 miles long and 50 miles wide, present at depths varying from a few inches to a few hundred feet, and for twenty years they offered the chance of fabulous riches to anybody who cared to come and dig. Never before or since in the annals of the American dream did its promise prove so demonstrably true.

The gold rush attracted expectant capitalists from everywhere in the world—not only other Americans from Ohio and Vermont but also Frenchmen, Chinese, Mexicans, Italians, Irishmen, Germans, Dutchmen, Swedes—all of them optimists, most of them young and male, few of them burdened with the luggage of civilization and its discontents. Because they arrived at more or less the same moment, they got off to a more or less even start in a new country unmolested by the nuisances of law, government, tradition, social custom, curfews, or prior claimants. San Francisco enlarged its population from 800 in 1848 to 30,000 by 1851, prompting Bayard Taylor, an early traveler in the city, to liken it to "the magic seed of the Indian juggler, which grew, bloomed, blossomed, and bore fruit before the eyes of his spectators."

Excused from a term of apprenticeship as a territory, California was admitted to the union in 1850, the resident aliens already in the habit of making up their own rules, owing nothing of their newfound circumstances to the existence of old ideas, settled monopolies, eastern money. The motley character of their society (plural, cosmopolitan, tolerant, and unstable) guaranteed the freedom of movement and encouraged, or at least didn't frown upon, the freedom of thought. Like the hero in a Clint Eastwood movie or a Raymond Chandler novel, the California protagonist belongs to no establishment, a born renegade fond of mocking the shabby masquerades (of traffic courts and dictionaries and jails) with which the corrupt officials in City Hall or Chinatown seek to imprison the noble savage dreamed of in the philosophy of Jean-Jacques Rousseau.

It is no accident that California over the last 150 years has furnished the country with so many of its new directions, most obviously in the entertainment business and the computer industry (a second manifestation of the miracle in the fortune-bearing sand) but also by way of its enthusiasm for Scientology, environmental ballot initiatives, sexual experiment, aircraft design, muscular artichokes blown up to the weight and size of Mr. Universe. Now as in 1848, the belief that wealth follows from a run of luck fosters among the Californians a willingness to deal the cards, take the chance, entertain the proposition offered by the gentleman wearing the mismatched boots or folded into the note from the lady with the parrot. Who knows? Maybe one of them will bring rain.

California is a desert, heavily populated by nomadic souls searching for the fountains of eternal youth, certain that if not the abundance of nature then the wonder of the consumer markets or the Indian juggler's magic seed will provide the people of the caravan with limitless pasturage for their camels and immediate seating at a poolside table in the lost oasis under Barbra Streisand's palm and eucalyptus trees. West of the Sierra Nevada, the delight in metaphor matches the wish to believe in what isn't there. Most of the people likely to be met with at a gas station in Bolinas or on the beach at Santa Monica turn out to be playing a part in a movie of their own invention. The reflections in the camera's eye seldom bear much resemblance to what an uninitiated bystander might mistake for reality, but to ask of the director too many disheartening questions, or to remember what he or she said yesterday or last week, constitutes an act of social aggression.

Who dares to speak to such people of death or law or taxes?

During the endless summer of 1978, the same year that Arnold Schwarzenegger made his fateful move toward the bikini wax, the children of paradise passed by an enthusiastic majority the grand remonstrance known as Proposition 13, which stripped the state government of its authority to set property taxes. The ballot initiative served as prologue to the reactionary triumphs of the Reagan revolution that now have brought to California the gifts of political bankruptcy and economic ruin. How better to express the abiding hatred of government, any government, than by sending to Sacramento a robot with a gun? Where else does California wish to go if not into the sunset or the sea with Conan the Barbarian?

So apparently does the rest of the country. Reading from the California script, Ross Perot announces his presidential candidacy on *Larry King Live*; the voters in Minnesota elect as their governor a professional wrestler; the state legislature in Texas shuts down operations because half of its members fled first to Oklahoma and then to New Mexico; Michael Bloomberg, a billionaire as poorly qualified for public office as Mr. Bush or Mr. Schwarzenegger, pays $60 million for the mayoralty of New York City and the pleasure of traipsing around a political soundstage in a movie of his own invention; in Washington the Bush Administration seeks to rid itself of nearly every government service (schools, prisons, electrical transmission lines, tax collections, broadcast frequencies) that it can sell or deliver to a church.

The high-end thinkers at the policy institutes in Washington and New York like to imagine the presence of such a thing as an American body politic, coherent and well informed, capable of carrying forward the work of a democratic republic. They ignore the fact that a good many of their compatriots have adopted the California frame of mind, wandering bedouin in place of settled citizens, drifting west and south toward the Garden of Eden in Las Vegas, many of the tribes illiterate or speaking strange languages, few of them familiar with the Fourth Amendment or the name of Patrick Henry. Twenty percent of the population shifts its household or habitation every year, and much of what now constitutes American society has a tentative and makeshift character, as if it had been put together for a season and was meant to be consumed at the point of sale. Even within the supposedly stable and institutional structures of the corporation, the Army, the media, the universities, the interior movement is as restless as water. Promotions fail to accrue to the accounts of people who stay too long in one place, who exhibit the virtues of patience, citizenship, loyalty, and honor. Money attaches itself to velocity, to the changing of occupations or employers at least once every six or seven years, and the country swarms with bands of itinerant opportunists forever keeping their "options open," conceiving of democracy as a pastoral romance.

The settled townsman makes art, science, government, and law; of necessity he comes to understand the obduracy of the soil or the stone, and he measures his victories over periods of time longer than those sold on television. The nomad merely gathers together his tent, his music, and his animals, and wanders over the mountain in search of next year's greening of America. By nature predatory and by temperament aristocratic (one who takes but doesn't build), the bedouin of the great American desert like to imagine themselves as mighty warriors seizing the spoils and emblems of their magnificence—oil companies, baseball teams, gold-plated golf clubs. Consumer markets welcome them as the golden horde that sustains the national economy with its voracious appetite and sets in motion the blessed chain reaction that leads to more spending, more investment, more employment, more highways, stronger steel, more confidence, more traffic deaths, more missiles, more amphetamines, more forms to fill out, more firepower, better cosmetics, more fear. Most prized of all the nomad's possessions is the heroic sense of self, and in place of the questions asked by citizens in a public square ("Who are they?" "What is this?"), the princes of the desert stare lovingly into the pools of Narcissus and ask—much to the joy and profit of their attendant dance instructors, psychotherapists, plastic surgeons, dealers in rare jewels and exotic religions, designers of designer jeans— "Who am I?"

Expressed in the acquisitive instead of the investigative case, the question becomes "What about me?" which was the form used by President Bush at a news conference in Crawford, Texas, on August 13, when he was surprised by the temerity of a reporter describing the entry of the actor Schwarzenegger into the California gubernatorial race as "the biggest political story in the country." The colloquy merits quotation in its entirety:

THE PRESIDENT (insulted): It is the biggest political story in the country? That's interesting. That says a lot. That speaks volumes.

FIRST REPORTER: . . . You don't agree?

THE PRESIDENT (irritated): I don't get to decide the biggest political story. You decide the biggest political story. But I find it interesting that that is the biggest political story in the country, as you just said.

SECOND REPORTER: You don't think it should be?

THE PRESIDENT (cute and sarcastic): Oh, I think there's maybe other political stories. Isn't there, like, a presidential race coming up?

In California the Hollywood press corps would have known better. It's never any good mentioning the names of two action-movie stars at the same news briefing. The technical assistants don't bring enough bikini wax for both heroes, and one of the seedy, freelance cameramen invariably makes a dumb mistake with the credit lines, forgetting whether it was Arnold Schwarzenegger in *Moon over Baghdad* who said, "Hasta la vista, baby!" to Saddam Hussein or George Bush in *California Sunshine* who said, "Bring 'em on" to the Sacramento chapter of Al Qaeda. By the time the publicists smooth things over with an agreement on the wording of the next day's gossip item, the star has stopped trying to answer questions in English.

October 2003

Revised Text

*When it's a choice between writing the story
and writing the myth, write the myth.*
—John Ford

During the week in early November when CBS canceled its broadcast of *The Reagans,* the gossip making the season's tour of the New York book-party circuit damned the network for stuffing a gag in the mouth of the First Amendment. The verdict was unanimous, announced by poets in Greenwich Village lofts, by publishers in offices overlooking Times Square, by novelists loose among the buffet tables at The Four Seasons and The Plaza Hotel. In every jurisdiction the finding was handed down with solemn turns of phrase: "dark night of fascism," "sad day for artistic freedom," "appalling cowardice."

No members of the literary jury had seen the four-hour movie that served as the occasion for their alarm, but then neither had the claque of self-appointed judges that forced CBS to remove the offending object from the display window of prime-time television. The complaints arose from rumors published in the *New York Times* on October 21, four weeks before the dramatization of Reagan's life and love story was scheduled to air. The producers were said to have intended a heartwarming biography—young actor rises to fame and fortune in Hollywood, finds God and an adoring wife, enters politics, moves upward to the White House, and saves the country from poverty and fools—but the screenwriters apparently had blotted their pages with a few careless lines of dialogue suggesting weevils in the ointment and flies in the milk. The first rumors—of scenes depicting Reagan as an intolerant homophobe guided by the wisdom of his wife's astrologer—were amplified by a second edition of rumors seeping into the Internet with the Drudge Reports of October 24 and 26—scenes depicting Reagan as a man who confused himself with the anti-Christ and Nancy as an angry dwarf screaming insults at the White House wine steward. It didn't matter that in the source materials the producers could show probable cause for the complications of character; nor did it matter that Reagan emerged at the end of four hours as an heroic president who restored America's faith in itself and won through to victory in the Cold War against the Russians. What mattered was the network's failure to deify Reagan. The man deserved an image made of gold; any hint of imperfection implied not only carelessness but also blasphemy.

The rumors of desecration were enough to summon from the Republican faithful a swarm of 80,000 outraged emails, the threat of boycott against automobile and soft-drink manufacturers associating their product with the movie, a demand from Ed Gillespie, chairman of the Republican National Committee, that the network submit the production to prior review by competent historians.

The thunder on the right was sufficient to its purpose, and on November 4, CBS canceled the broadcast of a drama on which it had spent $10 million, and had scheduled for Sweeps Week, during which the networks seek to attract their biggest audiences and highest ratings. The press release attributed the erasure to artistic integrity, a belated recognition on the part of the executive producers that *The Reagans* didn't meet the standard of "balanced portrayal" for which CBS was justly famous, the decision "based solely on our reaction to seeing the final film, not the controversy that erupted around a draft of the script."

None of the parties to the outcry credited the statement with even the semblance of truth. The friends of free speech in New York received it with scorn and derision; Ronald Reagan's champions in the suburbs of cyberspace accepted it as proof of the avarice and hypocrisy they always had known to be the earmark of the lying, liberal press. The CBS executives presumably didn't need to be reminded of their own Dan Rather's epitomization of network television as a natural-born toady. Rather had offered the analysis with reference to the false heroics of the reporting of the first American war against Iraq. Here in media fantasyland, he had said, "We begin to think less in terms of responsibility and integrity, which get you in trouble . . . and more in terms of power and money. . . . Suck-up coverage is in."

Suck-up coverage is always in, both in the print and the broadcast media, but if the disappearance of *The Reagans* would have come as no surprise to John Ford, the Hollywood director now best remembered for his mythological rendering of the old American West in the movie *Stagecoach,* it offered an instructive commentary on the revisions made over the last twenty years in the language of political correctness. Conceived in the universities in the early 1980s, the magical use of words was meant to empower what was then perceived as the disenfranchised left—women, gays, blacks, Latinos, environmentalists, any and all victims of circumstance. The speaking in euphemism, like the speaking in tongues in Pentecostal churches, supposedly conferred upon the devotees the mandate of

Heaven. Teach people to talk sweetly to one another, divide the curriculum into communities of uplifting sentiment (sexual and cultural as well as racial and sociological), and then surely, in the fullness of time and after many weavings of the sacred spells, they will learn to behave properly, to abandon the joys of date rape and haul down their Confederate flags, quit stealing from the poor to feed the rich, come to see and know that the true path to political happiness and constitutional redemption is to be found not with a compass or the 82nd Airborne Division but with a well-thumbed thesaurus.

The society's corporate managers needed twenty years to fully appreciate the nature of the gift they had been given. The jargons of safe speech struck them on first hearing as un-American and wrong, and they recruited several regiments of reactionary historians to wage what was billed as "the Culture War" against the barbarian hordes issuing forth from the English departments at Duke and Yale. The editors of the *Wall Street Journal* posted scouts on the perimeter of the defense budget to watch for feminists and deconstructionists; the neoconservative think tanks declared a state of emergency and set up assembly lines for the production of academic papers asserting that black people ask for nothing better than the comfort of continued failure; Rush Limbaugh faced down mobs of subversive adjectives; William Bennett sent up barrage balloons filled with helium and virtue.

The events of September 11, 2001, placed the doctrine of political correctness in a truer light and a clearer perspective. The distortions of language formerly regarded as the enemy of free expression stood revealed as the friend of Operation Enduring Freedom. With only slight adjustments in emphasis, the rinsed and blow-dried vocabulary proved better suited to the service of "power and money" than to the defense of "integrity and responsibility." The various commentaries on the omission of *The Reagans* in last November's *TV Guide* suggested notes toward a revised glossary.

I. APPROPRIATENESS   The universities employed the word to guard underprivileged individuals against attacks on their emotional well-being. Nobody knew how to change women into men or black people into white people, but the least that anybody could do was to pretend otherwise. It wasn't appropriate to observe that the progenitors of Western civilization were for the most part European, male, and white; it wasn't appropriate to say that the entire body of literature assembled on the whole of the African continent doesn't match, either in quality or in volume, the achievement of the city of London; it wasn't appropriate to say that women tend not to be drawn to the study of mathematics or drafted into the secondary of the Chicago Bears. The polite lies were meant to bind up wounded sensibilities, not to burnish the images of state.

In the interest of the latter purpose, the objectors to the Reagan mini-series borrowed the rules of etiquette made to the measure of the former. It wasn't appropriate to speak poorly of a man afflicted with Alzheimer's disease, "a little tacky," according to *Newsweek,* "to be taking a lot of pot shots" at a president unable to come to a microphone, "in questionable taste," according to the *Washington Post,* "to schedule a controversial drama" apt to hurt the feelings of so many fine and loyal Americans. Whether the movie was improbably good or predictably bad mattered as little as a critical evaluation of Maya Angelou's poetry.

Deployed in the theaters of military operation, the polite uses of language allow our military spokespeople to conduct their briefings in the manner of English professors teaching a class in semantics. Nobody knows how to fight wars without killing people, but the least they can do is pretend otherwise. Thus Lieutenant General Ricardo Sanchez, at a press conference in Baghdad on November 11, announcing the harsh disciplinary action (Operation Iron Hammer) soon to be meted out to Iraqi guerrillas: "Although the coalition can be benevolent, this is the same lethal instrument that removed

the previous regime, and we will not hesitate to employ the appropriate levels of combat power."

None of the reporters attending the class inquired about the
chance of casualties—military, civilian, or collateral. To do so
would have been "inappropriate," for the same reasons that the
Bush Administration deems it "inappropriate" to permit photographs of the flag-draped coffins off-loaded on the ramps at Dover
Air Force Base.

2. AUTHENTICITY   By 1990 at many of the nation's right-thinking
universities, the novels of Jane Austen and George Eliot had been
remanded to the custody of the department of Women's Studies,
the texts subject to explication only by female professors of literature. A majority of the critics complaining about *The Reagans* followed a parallel line of argument to their opinion of its
worthlessness. The actor James Brolin had been hired to play the
part of the former president, and because Brolin is married to Barbra Streisand—notorious liberal, friend to Bill Clinton, hater of
Republicans—Brolin clearly came to the set with vicious intent.
How could it not be so? The man was marked by the company he
kept, and "what cruelty must lie in the hearts of CBS executives" to
sit a Communist on the Gipper's horse?

3. HATE SPEECH   An all-purpose synonym for bigotry, favored by
college deans in the 1980s as a caution against ugly insults directed
at a student's sexual or racial orientation. The phrase lately has come
into more general usage as a restraining order on expressions of disagreement with American foreign policy. The Pentagon's cadre of
high-ranking geopolitical strategists consists largely of fierce and visionary ideologues who advocate the social and political restructur-
ing of the entire Middle East—Islamic militancy suppressed, Arab
nationalism uprooted, Palestinian radicalism destroyed. They make
little distinction between the objectives of the Bush Administration

in Washington and the Sharon government in Jerusalem, but requests for further clarification invite the charge of anti-Semitism.
Jews who ask questions find themselves modified by the adjective
"self-hating." When George Soros on November 5 informed an audience of Jewish philanthropists in New York that in Europe "there
is a resurgence of anti-Semitism" and that the policies of the Bush
Administration "contribute to that," his remarks were denounced,
within a matter of days, by Congressman Eliot Engel (D., N.Y.) as
"ridiculous and outrageous," "morally reprehensible."

5. RACISM   Let it once be established that prejudice is an evil with
a thousand faces and as many names (the bias against fat people or
Yorkshire terriers as small-minded and contemptible as the bias
against black people or parakeets), and there's no end of the services
that the word can be made to perform. The intelligentsia on the
Republican right picked up on the possibilities during the first
term of the Clinton Administration, developing the theory of rich,
white men as an oppressed minority suffering under the lash of federal tax policy and Michael Moore's jokes. The election of George
Bush brought them out of the closet. Having come to see their selfishness as a form of ethnic identity too long submerged by an alien
culture insensitive to the humiliation of a Wednesday without oys-
ters or a Sunday without golf, they have found in Grover Norquist,
president of Americans for Tax Reform, their own Martin Luther
King, leading them out of darkness, striking off their chains. Not a
man afraid to speak out against social injustice, Norquist entered
the plea of racism on October 2, in a conversation with Terry Gross
on the National Public Radio program *Fresh Air*. Comparing the
cadre of wealthy Americans to the victims of the Holocaust,
Norquist said,

> The morality that says it's okay to do something to a group
> because they're a small percentage of the population is the

morality that says that the Holocaust is okay because they didn't target everybody, just a small percentage. . . . Arguing that it's okay to loot some group because it's them, or kill some group because it's them, and because it's a small number—that has no place in a democratic society.

6. CULTURAL HERITAGE   Some cultures discover their identity in clouds or the humming of bees; other cultures find coherence in metaphysics, their traditions shaped by the five hundred words for snow or the ninety-nine names of God. Our own culture dances to the music of its guns. Warfaring people, unique in our gift for violence, our best-loved myth is the one about the going west with the rules of forward deterrence and preemptive strike, killing anything and everything (the buffalo and the Indian, the beaver, the top soil, the passenger pigeon, and the Mexican) standing in the way of progress on the long and lonesome trail to Rancho del Cielo.

For too many years we have denied our heritage, been made to feel embarrassed by the bombings of Hiroshima and Dresden, forgotten that the Republican Party won the Civil War, shied away from the wisdom of William Tecumseh Sherman, who knew and said that "we must act with vindictive earnestness against the Sioux, even to their extermination, men, women, and children."

Fortunately we have the war in Iraq to recall us to a sense of who we are; fortunately we have not only President George W. Bush but also the *New York Times* columnist David Brooks to set the proper tone. Thus Brooks on November 4, saying that true Americans don't flinch in the face of death appropriately administered to people whom we neither see nor know:

What will happen to the national mood when the news programs start broadcasting images of the brutal measures our own troops will have to adopt? Inevitably, there will be atrocities that will cause many good-hearted people to defect from the cause. . . . The President will have to remind us that we live in a fallen world, that we have to take morally hazardous action if we are to defeat the killers who confront us. It is our responsibility to not walk away.

The accommodating moralist reconfigures the code of etiquette, which once forbade the wounding with injurious words, to excuse the committing of inevitable atrocities; the objection to "brutal measures" is reperceived as an action "morally hazardous." The revisions should come as no surprise. The labels of political correctness never were intended to convey meaning; analogous to a notary's stamp, they were meant to certify the transfers of jurisdiction and propriety, instruments of power instead of thought. The freedoms of expression prove contingent upon the circumstances, the instruction on the labels changed to match the preferences in virtual reality.

January 2004

Bad Medicine

Some circumstantial evidence is very strong, as when you find a trout in the milk.
—Henry David Thoreau

*F*orty years ago I was not yet thirty, and my father still held to the hope that I would come to my senses, abandon the practice of journalism, and follow a career in one of the Wall Street money trades. As a young man during the Great Depression he had labored briefly as a city-room reporter for William Randolph Hearst's *San Francisco Examiner,* and he knew that the game was poorly paid and usually rigged, more often than not a matter of converting specious rumor into dubious fact. Having escaped the sorrows of Grub Street and gone east to become an eminently respectable New York banker, also the director of Fortune 500 companies and a member of long-established clubs, he understood the principle on which the society arranges its socioeconomic seating plan, and for his son he wished a place on the dais, preferably with a view of both the mountains and the sea.

The lesson was not one that my father knew how to teach in

words of his own choosing—possibly because he believed, and couldn't avoid saying, that even the most successful journalists sit well below the salt—and so he made a point of introducing me to various captains of industry and finance in whose lives and works I might discover the happy result of a properly schooled ambition. Over a period of three or four years in the middle 1960s I found myself in golf foursomes with oil-company presidents, at lunch or dinner with the senior partner of an investment bank, present in a box at Yankee Stadium with the managing director of a trust or a pension fund. None of the instruction led to the desired consequence, but among all the fond reflections on fortunes won or lost, I remember best the cautionary tale told by Theodore Weicker while walking on a beach at East Hampton.

A man then in his late sixties, as rich and well-groomed as he was wise in the ways of the world, Weicker had inherited, together with his brother, the controlling interest in the Squibb pharmaceutical company, and shortly after the Second World War, divining the possibilities implicit in what subsequently has come to be known as "globalization," he traveled to the poorer countries of the earth to set up factories producing drugs for the local pharmacies and hospitals. He learned early in the proceedings that it was better to deal with dictators than with democracies. Doing business in a democracy necessitates the bribing of too many people, most of them more than once. "Nobody stays bought," he said. "When anything goes wrong, somebody to whom you just paid $10,000 stands up in whatever they call a congress in those places and makes a speech about American exploitation."

The Shah of Iran appreciated the difficulties, and by way of illustration, Weicker told the story of the Iranian health minister who objected, "on humanitarian grounds, no less," to a widening of the profit margin on the manufacture of aspirin. For some years Squibb and the Iranian government each had been content with markups of 100 percent over cost, but then it occurred to the government to rub Aladdin's lamp for another 100 percent. The health minister refused to approve the request, and for a period of several months nothing could be done. The chances for negotiation improved when the health minister checked into a hospital for a routine appendectomy. As soon as he had been settled on the operating table, the nurses informed him that the hospital had exhausted its supplies of anesthetic. The surgeon apologized for the inconvenience while holding a scalpel to the minister's shaved and trembling belly, explaining further that the pain of incision would be so great as to deliver the minister into a state of shock, which, although momentarily unpleasant, would serve as a kind of anesthetic. The minister consented to the rise in price.

Weicker delighted in the elegant simplicity of the transaction. No lawyers, no environmental-impact statement, no waiting around for a report from the ethics committee. The true nature of the free market revealed with a clarity and precision seldom attained by the professors of economics at Harvard or Yale. "There you have it," he said. "What business is all about. Building a better world."

The cautionary and uplifting tale came to mind in early December last year when President Bush signed the amendment to the Medicare legislation that delivers 40 million elderly and disabled American citizens into the custody of the good-hands people operating the nation's insurance and pharmaceutical factories. The new authorization purports to reduce the cost of prescription drugs for every needful American, and the White House staff dressed up the photo op in Washington's Constitution Hall to look like a scene of joyful thanksgiving—a vast throng of well-wishers, military band music, a bright blue banner emblazoned with the physician's comforting "Rx," grateful invalids and smiling congresswomen, President Bush in the part of a merry Santa Claus bestowing upon the multitude the gifts of Christmas yet-to-come. Hats off gentlemen, send for the champagne. Great, good news; bread upon the waters; pennies raining from heaven and stars falling on Alabama.

Looked at a little more closely, the scene acquired a somewhat different character and tone. Still celebratory and festive, of course, but the rejoicing of bandits and thieves as opposed to the thankfulness of survivors rescued from a shipwreck. It was hard not to think of Eskimos contemplating the bonanza of a beached whale, the faces in the crowd those of K Street lobbyists eager to congratulate the politicians (chief among them J. Dennis Hastert, speaker of the House of Representatives, and Dr. Bill Frist, the Senate majority leader) who had worked so long and hard to unblock the river of government money now free to water the plains of avarice. It was the genius of Hastert that had formulated the legislation in 681 pages of stupefying prose (and strong-armed the rules of parliamentary procedure in the House to secure the winning vote at 6:00 A.M. on November 22), and it was the calm and morally anesthetized composure of Frist that in the Senate on November 25 had placed the scalpel of extortion against the shaved and naked flesh of the American body politic.

Few or none of the politicians who voted either for or against the bill took the trouble to read it; like them, I rely for my understanding of it on what I've seen in the newspapers and what I've been told by informed medical practitioners, but I think it safe to assume that the particulars speak to Weicker's ideal of free-market perfection. The principal author of the legislation, Thomas A. Scully, set about the task of writing it in June of last year, while he was employed as the federal administrator of Medicare. At the same time he expressed the wish to enter the private sector, putting his services up for auction to five high-priced Washington influence brokers representing the insurance companies, the drug manufacturers, and the health-maintenance organizations. Eight days after the happiness in Constitution Hall, Scully resigned his government post to await bids for his tour guide's knowledge of the small print that allots as little money as possible to individual citizens and as much money as possible to the vested commercial interests.

Although the government must provide drugs to 40 million people, it may not negotiate a bulk discount; it must pay whatever price the manufacturer sets or asks—prices that in the recent past have been rising at a rate of 12 percent a year. The legislation forbids the importing of less expensive drugs from Canada, prohibits beneficiaries from buying supplemental insurance for drugs unacknowledged by Medicare, reduces or eliminates payments to as many as 6 million people for whom Medicaid now defrays at least some of their prescription costs, declares a suspension of payment at precisely the point when most people might need the most help. An annual premium of $420 covers 75 percent of drug expenditures up to $2,250; from that point upward the beneficiary must pay, with his or her own money, 100 percent of the next $3,600 in costs; once the expenditures reach a total of $5,850, the government pays 95 percent of the subsequent bill. The actuarial tables assume that relatively few people can afford (or will live long enough) to pay the toll on the bridge across the river of public money flowing out of Washington into the privately owned catch basins of the medical-industrial complex. As a further means of implementing the shift of the nation's health-care burden from the public to the private sector, the legislation offers various inducements to the life-enhancing profit motive:

A. A $12 billion slush fund from which, over the next ten years, the secretary of health and human services may pay out bribes to HMOs otherwise reluctant to accept patients whose illnesses cannot be prepped for a quick and certain gain.

B. A windfall of $70 billion, also to be provided over the next ten years, to those corporations willing to continue prescription-drug coverage for their retired employees, the money to be paid in the form of both tax deductions and tax-free subsidies.

C. The guarantee of "maximum flexibility" to the private entities seeking to recruit customers from the general population now served by Medicare. The private entity may exercise the right to "cherry pick"—i.e., to offer its services only to those individuals not likely to require expensive treatment. The government must provide for everybody else, for the hopelessly enfeebled and the terminally indigent.

D. The legislation's reliance on the drug companies and the private insurers to curtail spending and control costs. The provision serves a dual purpose. It assures the eventual destruction of the entire Medicare apparatus, and it relieves the government of any responsibility for what will be reported as an act of God. Even the dimmest of Republican congressmen knows that the government doesn't have the $400 billion that the drug-prescription benefit presumably will cost over the next ten years—doesn't have the cash on hand or anywhere in anybody's budget projection. The money must be borrowed, at rates of interest yet to be determined. In the meantime, while waiting upon possibly unhappy financial events (wars, revenue shortfalls, stock-market downturns, sustained recession, etc.) the government retains no control of the fees charged by the health plans or the prices that the pharmaceutical companies demand for drugs. Let Mother Nature take her course, and the expenditure estimated at $400 billion easily could become an invoice presented for $1 trillion.

Among all the cheats and false suppositions written into the legislation, the assumption that private entities somehow might be induced to restrain spending should have been the easiest one to ferret out, if by nobody else than by Bill Frist, the Senate majority leader. The good doctor knows, probably better than any of his colleagues in the Senate and certainly as well as Ted Weicker's exemplary surgeon in long-ago Teheran, how to inflate a drug cost, supply an unnecessary medical procedure, pad a hospital bill. In 1968, Frist's father and elder brother established the Hospital Corporation of America (HCA), which has since become the nation's largest consortium of for-profit hospitals and medical centers—252 of them in twenty-three states. For several decades the company required each of its hospitals to return a profit of 20 percent a year and to "upcode" their patients by exaggerating the degree and severity of their illnesses in order to receive, from Medicare, more generous reimbursements for the delivery of imaginary goods and services. So skilled did the hospitals become in the arts of medical chicane that in December 2000 HCA admitted to a defrauding of the federal government so massive that it required the payment of fines in the amount of $840 million. Two years later, confronted with a supplementary set of similar charges, the company negotiated a settlement for an additional $631 million. The agreement was reached on December 18, 2002, two days before Frist was elected Senate majority leader.

Another cautionary tale, but not one supportive of the hope that the cost of the prescription-drug benefit will be contained by the people dispensing it. The corporate health-care systems that currently hold captive 160 million Americans (in return for an annual ransom of $952 billion) can't afford the luxury of a conscience or a heart. They're set up to make money, not to care for sick people, and even if the managers sometimes might wish it otherwise, how then would they pay themselves life-enhancing salaries, and what might happen to their faith in the free market? Before investing in private health-care organizations, the Wall Street financial analysts like to see a low "medical-loss ratio" (i.e., that percentage of the yearly revenues actually allotted to patient care) sufficient to offset the administrative costs (9.5 percent in the private sector as opposed to 1.4 percent in the public sector) as well as fund the annual compensations awarded to the chief executives—an average of $15.1 million in 2002 at the country's eleven leading health-care companies. Even in the best of circumstances the miracle of the free market is never easy to maintain, but over the last few years healthy numbers have become more difficult to find, and if not to their friends in the Congress and the White House to whom else does a good American turn for a little help with the building of a better world?

Two days after the House of Representatives passed the legislation by a vote of 220 to 215, Speaker Hastert's spokesman named the reward expected in return for so handsome an act of friendship. "This is the thing," he said, "[Hastert] thinks will keep us in the majority for a while." Not forever, not after the legislation takes effect in 2006, but at least until next November's elections, for as long as the specious promise can be promoted as an authentic fact and before too many people open their Christmas presents to find nothing in the box except a card wishing them a happy New Year and hoping that they get well soon.

Proud of its plundering of the American commonwealth on behalf of its corporate sponsors and political accomplices, the Bush Administration follows a practice well established by both its near and distant predecessors. The raids on the federal treasury encouraged by the Reagan Administration took place under the cover of a darkness represented as ideological enlightenment. Deregulation was the watchword for the transfer of wealth from the public to the private sector, the $500 billion savings-and-loan swindle an exemplary proof of what could be done with the theory that big government (by definition wasteful and incompetent) deserved to be sold for scrap to the entrepreneurs in our midst (by definition innovative and efficient) who know how to privatize the profits while socializing the risk and the cost. Wonderful news, said the *Wall Street Journal,* pennies falling from heaven and stars on Alabama, more swill for the pigs. Diligently applied by a succession of industrious thieves over the last twenty-five years, the theory has resulted in the wreckage of the deregulated airlines, the degradation of the environment, the monopolies strangling the wit and sense out of the news media, the Enron debacle, most recently the Halliburton company's theft of $61 million (configured as a 100 percent markup on the price of gasoline) from the American army in Iraq.

The looting is traditional, the rule of capture as firmly rooted in the country's history as the belief in angels. Emblazoned with the mottos "Boom and Bust," "Settle and Sell," "Get In, Get Rich, Get Out," the winning of the nineteenth-century American West was a public-works project paid for with federal subsidy. By 1850 everybody traveling west of the Mississippi understood that the country was ripe with four primary resources—land, minerals, timber, and the government contract—and that of these, by far the richest was the government contract. The trick was to know the right people in Washington, at the state capitol, and at the county courthouse. When the two sections of the transcontinental railroad were joined with a golden spike in May of 1869 at Promontory Point, Utah, the patriotic ceremony in the desert (band music, red, white, and blue bunting, hats in the air) glossed over the swindling mechanics of the prototypical government cost overrun. The work was so shabby that much of it had to be replaced within a year, the railroad setting up dummy corporations that rigged the prices of reconstruction, and the bipartisan majority in Congress content to sell its ethical interest for a percentage of the gross.

The Medicare drug benefit fits the job description understood not only by the treasury officials in the Grant Administration but also by the ambulance drivers who picked up Andy Warhol on the summer day in 1968 when he was shot twice in the stomach by a deranged movie actress. The drivers found Warhol on a dingy street in SoHo, and on the way to the hospital, not recognizing him as a celebrity and thinking him as worthless as everything else in a neighborhood not yet trendy, they didn't feel compelled to hurry.

Warhol noticed that the ambulance was stopping for red lights and mumbled something about the urgency of his wound. The drivers looked at him as carefully as an HMO accountant looks at the pre-authorization request for a mammogram or a crutch. "For fifteen bucks," one of them said, "we'll turn on the siren."

February 2004

Dar al-Harb

Men never do evil so fully and so happily as when they do it for conscience's sake.
—Pascal

*D*uring the two and a half years since the terrorist attacks on
New York and Washington, the country's book publishers
have poured forth a steady flow of propaganda recruiting the American citizenry to never-ending war against all the world's evildoers.
The edifying tracts come in two coinages—those that praise America the Beautiful (virtuous and just, forever innocent and pure in
heart) and those that magnify the threat posed by sinister enemies
as numberless as the names for grief—nuclear weapons in the
hands of North Korean generals and Pakistani bicyclists, smallpox
virus hidden in the luggage or on the person of a gentleman from
Bolivia, Arab fanatics spawning in the sewers of the once romantic
Middle East. The sales pitch can be inferred from a short but representative list of titles—*Taking America Back; Why We Fight; Ripples
of Battle; An Autumn of War; A Heart, a Cross, and a Flag.* I can't pretend to having read more than a few of the hundred-odd books

made to the design specifications of a Pentagon press release, but judging by those of which I've read at least enough to appreciate the author's command of the false but stouthearted syllogism, the attempt at persuasion appears to have shifted from the secular to the religious lines of argument.

The books published during the first twelve months of the country's introduction to the concept of its own mortality stated the problem as one open to solution with the instrument of reason, possibly also with some knowledge of history and a passing acquaintance with the socioeconomic circumstances confronted by a majority of most of the world's peoples, Asian, African, and Latin American as well as Muslim. The writing wasn't distinguished, but at least it could be said that the discussion was taking place in the vernacular languages of the world in time. The more recent books borrow their inspiration from the verses of the Bible and the suras of the Koran. The authors who decry the sins committed by Americans in America (cynicism, homosexuality, believing what they read in the *New York Times*) adopt the rhetoric of Jonathan Edwards cleansing the souls of the unfaithful in the seventeenth-century wilderness of Puritan New England; the authors who preach holy crusade against the foreign infidel in modern-day Jerusalem and Damascus issue fatwas in the manner of Osama bin Laden.

To the latter company of vengeful imams we now can add the names of David Frum, a former speechwriter for President George W. Bush, and Richard Perle, member and former chairman of the Defense Policy Board within the U.S. Department of Defense. Their jointly assembled "manual for victory" (*An End to Evil: How to Win the War on Terror*) reached the bookstores in early January and was promptly boosted onto the bestseller list because of the authors' elevated rank within the intellectual apparat that supplies the Bush Administration with its delusions of moral grandeur. Attached to the White House staff in 2002, Frum brightened that year's State of the Union Address with the phrase "axis of evil"; in

2003 he published *The Right Man*, a hagiography portraying President Bush as a man impervious to doubt, casting his mission and that of his country in the grand vision of God's master plan. Perle has served as an apostle of hard-line power politics since the early 1980s: an assistant secretary of defense in the Reagan Administration, closely associated with the troop of visionary ideologues currently directing the course of our militant foreign policy, a resident fellow, together with Frum, of the American Enterprise Institute.

The result of their collaboration is an ugly harangue that if translated into Arabic and reconfigured with a few changes of word and emphasis (the objects of fear and loathing identified as America and Israel in place of Saudi Arabia and the United Nations) might serve as a lesson taught to a class of eager jihadists at a madrasa in Kandahar. The book's title testifies both to the absurdity of its premise and the ignorance of its authors. Evil is a story to which not even Billy Graham can write an end; nor can the 101st Airborne Division set up a secure perimeter around the sin of pride. The War on Terror is a war against an abstract noun, as unwinnable as the wars on hunger, drugs, crime, and human nature, and were it not for the authors' involvement with the affairs of state (informed sources, highly placed, presumably knowing why and whereof they speak), the entire press run of their book might be more usefully transformed into a shipment of paper hats.

But ours is an age in which hijacked airliners and precision-guided cruise missiles follow flight paths dreamed of in the minds of men like mullahs Omar, Frum, and Perle, and some of the prospective casualties (most of them civilian) might care to read at least a brief summary of the work in hand. As with all forms of propaganda, the prose style doesn't warrant extensive quotation, but I don't do the authors a disservice by reducing their message to a series of divine commandments. Like Muhammad bringing the word of Allah to the widow Khadija and to the worshippers at the well Zem-Zem, they aspire to a tone of voice appropriate to a book of Revelation.

THE WAR ON TERROR—"Has only just begun." "There is no middle way for Americans: It is victory or holocaust."

THE UNITED STATES—"The greatest of all great powers in world history."

MILITANT ISLAM—Vast horde of Arabs, possibly numbering in the millions, bent on world domination. They intend to "fasten unthinking, unquestioning slavery" on the mind of Western civilization. As insatiable in their thirst for blood as Adolf Hitler and Joseph Stalin.

THE CIA AND THE FBI—Staffed by "faint-hearts" who lack "the nerve for the fight."

THE STATE DEPARTMENT—"An obstacle to victory." Controlled by "complacent" functionaries who seek "to reconcile the irreconcilable, to negotiate the unnegotiable, and to appease the unappeasable."

THE UNITED NATIONS—"Not an entirely useless organization," but one that does more harm than good. It "regularly broadcasts a spectacle as dishonest and morally deadening as a Stalinist show trial, a televised ritual of condemnation that enflames hatreds and sustains quarrels that might otherwise fade away." The United States doesn't require the U.N.'s permission to attack any country in which George Bush, like Teddy Roosevelt before him, notices "a general loosening of the ties of civilized society."

THE MIDDLE EAST—"Fetid swamp" crawling with "venomous vermin." America must cast the whole of it into a purifying fire. "The toughest line is the safest line."

THE AMERICAN OCCUPATION OF IRAQ—A triumph. Another six months, and nobody will be able to tell the difference between Baghdad and Orlando.

SAUDI ARABIA—Headquarters tent of militant Islam. "The Saudis qualify for their own membership in the Axis of Evil." Must be purged of ideological infection.

ARABS RESIDENT IN PALESTINE—Must learn "to swallow their

defeat." They have as little chance of territorial restoration as the Oglala Sioux who once drifted through the valley of the Little Big Horn.

ISRAEL—Land of heroes, light unto the gentiles. "Everything that liberal Europeans think a state should not be: proudly nationalist, supremely confident, willing and able to use force to defend itself—alone if need be."

IRAN—"The regime must go."

CRITICS OF AMERICA'S WAR ON TERROR—Cowards, "softliners," "Clintonites," "defeatists," pillars of salt.

GERMANY—"Fair-weather friend." An ally no longer to be trusted.

FRANCE—An unfriendly power that "gleefully smashed up an alliance [NATO] that had kept the peace of the world for half a century." Jealous and resentful of America's greatness and goodness of heart. Europe must be forced "to choose between Paris and Washington."

NON-DEMOCRATIC GOVERNMENTS—America not obliged to honor their pretensions. "It's not always in our power to do anything about such criminals, nor is it always in our interests, but when it is in our power and our interests we should toss dictators aside with no more compunction than a police sharpshooter feels when he downs a hostage-taker."

DARK PLACES OF THE EARTH—Any city, town, desert, parliament, or shopping mall where "terrorists skulk and hide." Among the countries eligible for some form of instructive intervention, the authors list Lebanon, Sierra Leone, Colombia, Venezuela, Paraguay, Brazil, Zimbabwe, Syria, Yemen, Somalia, northern Nigeria.

NATIONAL IDENTIFICATION CARD—Necessary precaution. All Americans must learn to inform on one another, to admire the vigilant citizen "who *does his or her part.*" "A free society is not an unpoliced society. A free society is a self-policed society."

AMERICAN CIVILIAN POPULATION—Must be prepared to accept losses. The protection of American citizens not as important as the killing of Satanic Muslims. The authors cite General George Patton: "Nobody ever won a war by caring for his wounded. He won by making the other poor SOB care for *his* wounded."

In their setting forth of the reforms that must soon descend on Saudi Arabia (on pain of swift punishment if the reforms don't come soon enough), Mullah Frum and Mufti Perle complain of the puritanical clerics in that country who "preach hate-filled sermons, teach the most frightful lies, and disseminate the deadliest conspiracy theories," who claim the authority of God for exhorting the faithful to acts of "cruelty and evil." They describe themselves. Their book summons all loyal and true Americans to the glory of jihad, and as an indication of the sort of thing they have in mind they rely on the wisdom of Sheikh Yousuf al-Qaradawi, dean of Islamic studies at the University of Qatar and a fond admirer of the terrorist bombings in Jerusalem and Tel Aviv. The two Washington ayatollahs mean to damn the sheikh, not to praise him, but when they quote his interpretation of Dar al-Harb (the Domain of Disbelief) they define the concept that they have made their own. The sheikh put it as plainly as possible to a crowd of Western journalists in Stockholm in the summer of 2003: "It has been determined by Islamic law that the blood and property of people of Dar al-Harb (i.e., non-Muslims) is not protected. Because they fight against and are hostile toward the Muslims, they annulled the protection of his blood and his property."

Nor did the sheikh draw an unnecessarily fine distinction (defeatist, appeasing, faint-hearted) between the military and civilian targets of jihad. A good Christian is a dead Christian. "In modern war," so said the sheikh, "all of society, with all its classes and ethnic groups, is mobilized to participate in the war, to aid its continuation, and to provide it with the material and human fuel required for it to assure the victory of the state fighting its enemies. Every citizen in society must take upon himself a role in the effort to provide for the battle. The entire domestic front, including professionals, laborers, and industrialists, stands behind the fighting army, even if it does not bear arms."

The Bush Administration's senior geopoliticians, among them Tom Ridge at the Department of Homeland Security and Donald Rumsfeld at the Pentagon, give weekly press conferences at which they make more or less the same speech—our army in Iraq bringing the great truths of democracy and global capitalism to the Domain of Disbelief, our civilian population girded for battle at security checkpoints on every major highway, at every airport, river crossing, bus and subway stop. Last year on Christmas Eve the FBI issued a bulletin advising 18,000 of the country's law-enforcement agencies to watch out for people carrying almanacs—i.e., works that answer to the uses of the intellect rather than to the joys of superstition. Almanacs, said the FBI, can be used by terrorists "to assist with target selection and preoperational planning." Because almanacs contain information, often accompanied by photographs and maps, about waterways, bridges, dams, reservoirs, tunnels, buildings, and roads, anyone carrying such a thing might be a terrorist or a friend of a terrorist. It was suggested that police officers approach with caution all almanacs "annotated in suspicious ways."

Never having heard the angry teaching in a madrasa or a mosque, I can't make a close or fair comparison to the briefing papers passed around the conference tables at the American Enterprise Institute, but if what I've been told is true (that the sermons depend for their effect on the expression of high-pitched rage) then I don't know how the language differs from that of Mufti Frum and Mullah Perle. Their truths are absolute, their verbs invariably violent—"destroy," "smash," "purge," "deny," "punish," "cut off," "stomp." Provide them with a beard, a turban, and a copy of the Koran, and I expect that they wouldn't have much trouble stoning to death a woman discovered in adultery with a cameraman from CBS News.

Set aside the question as to whether *An End to Evil* proceeds from a cynical motive (the authors fully aware of the lies told to promote a fanciful dream of paradise), and we're still left with a frightening display of ignorance that doesn't augur well for the future of the American Republic. In place of reasoned argument, we have Stone Age incantation, the sense or knowledge of history grotesquely distorted in the fun-house mirrors of ideological certainty, the observations of two respected and supposedly well-informed civil servants framed in a vocabulary as primitive as the one that informs the radio broadcasts of Rush Limbaugh, the television commentary of Sean Hannity and Bill O'Reilly, the novels of Tim LaHaye and Saddam Hussein.

Historians who study the rise and fall of nations mark the downward turn at the point when the rulers of the state begin to lose faith in the merely human institutions that embody a society's courage of mind and rule of law. They place their trust in miracles and look for their salvation to charlatans who come to comfort them with stories about the end of evil. When the Turks sacked Constantinople in 1453, they found 10,000 people in the church of Santa Sophia, earnestly praying for deliverance in a sanctuary made sweet with the smell of incense and stale with the scent of fear. Authors Frum and Perle trade in the same commodities.

March 2004

The Fog of Self

If vanity does not entirely overthrow the virtues, at least it makes them all totter.
—La Rochefoucauld

The NBC program *Meet the Press* devoted its broadcast on February 8 to an hour-long interview with President George W. Bush, and well before Tim Russert interrupted the proceedings for the first commercial break, it was apparent that the White House publicists had been forced to desperate measures. The President was rapidly losing altitude in the opinion polls, no weapons of mass destruction had been found anywhere east of Suez, and the news media were uncharacteristically loud in their suspicion that our armies had gone off to war under false pretenses, that Saddam Hussein never had confronted the United States with a clear or present danger, that the threat had been fabricated (if not by the intelligence agencies then by the visionary ideologues at the Pentagon), and that 3,588 American soldiers were dead or wounded in Iraq not for God, for country, and for Yale, but for reasons yet to be announced.

Compound the bad news from abroad with the increasingly

obvious bias of the government's economic policies at home (money for the rich, unemployment for the poor, the federal debt rising to $521 billion in 2004), and the omens didn't favor the President's chance of reelection in November. How then to refurbish the illusion of the President's integrity? In the absence of a better option (posing their man on the deck of another aircraft carrier or sending him back to Baghdad to distribute Easter eggs) the Oval Office stage crew elected to prop him up in front of a network camera without a flight suit or a script. The mistake was evident in the first five minutes of the broadcast. Disappointing his admirers among the friends of Jesus and ExxonMobil, the President took it into his head to play at being clever—alert to the tricks of his adversaries in Congress and the liberal press ("It's tough here in Washington"), well satisfied with his persona as the lionhearted defender of women and children and Western civilization ("This is a dangerous world, I wish it wasn't"), smug in the assumption that with his lifetime guarantee of schoolboy charm he had proved himself a match not only for a world-class "madman" but also for the host of demonic assassins concealed in the labyrinth of "shadowy terrorist networks."

The interview was an embarrassment. The President isn't good with words, and he seldom knows how or why or when he's lying. Not having a straight story to tell, and unwilling to hold himself accountable to anything other than his own courage and resolve, he answered Russert's questions with statements deployed as evasive decoys similar to the chaff off-loaded by an F-16 dodging enemy ground fire.

"So, we need a good intelligence system. We need really good intelligence."

"I make decisions here in the Oval Office in foreign policy matters with war on my mind."

"And the American people need to know they got a president who sees the world the way it is."

"Saddam Hussein was dangerous with weapons. Saddam Hussein was dangerous with the ability to make weapons. He was a dangerous man in the dangerous part of the world."

"When the United States says there will be serious consequences, and if there isn't serious consequences, it creates adverse consequences."

"This economy has been through a lot, which is why I'm so optimistic about the future because I know what we have been through. . . . I think this economy is coming around just right, frankly."

Most of what the President had to say can be so easily refuted or disproved, if not with reference to the public record then on the evidence of several recently published books (among them *American Dynasty* by Kevin Phillips, Ron Suskind's *The Price of Loyalty*, Molly Ivins's *Bushwhacked*, and *The Lies of George W. Bush* by David Corn), that I don't think it necessary to argue with each of the already discredited assertions. A few of them serve to make the point: that the war on Iraq was forced on the United States by Saddam Hussein; that the war is in no way "political"; that Saddam possessed a vast store of monstrous weapons; that the Bush Administration's budgets are not constructed with false numbers and fraudulent projections; that America's word is now credible in the world because everybody knows that we mean what we say; that the commission appointed by the President to assess the country's intelligence capabilities surely will come forth, four months after the November election, with an honest accounting. Senator John McCain (R., Ariz.) deconstructed the last-named assurance on the day that he was appointed to the commission in question. Before examining a single witness or document, he said: "The President of the United States, I believe, would not manipulate any kind of information for political gain or otherwise."

If by now we know little else about the President of the United States, we know that no matter what the subject under discussion— Iraq, the budget, the environment, the distribution of tax refunds— the White House manipulates every phrase of every speech and press release to no other purpose except that of political gain. But if the substance of what the President said on *Meet the Press* contributed nothing to the inventory of general disinformation, the form and manner of the interview added to the common store of what John Adams named as "that most dreaded and envied kind of knowledge" available to a free people, "I mean, of the character and conduct of their rulers."

Among students of the Washington talk-show circuit Russert enjoys the reputation of a prosecutor so fierce in his cross-examinations that politicians supposedly tremble when crossing the threshold of his studio, afraid that he will expose them as cheats and liars. On the showing of his conversation with President Bush, the reputation is undeserved. The questions were polite, few of them followed with the indignity of a request for further clarification, and it was impossible to escape the impression of a prep-school headmaster listening to the richest boy in the senior class explain how and why he had burned down the library and the gymnasium. Both parties to the interview understood that the boy wasn't going to be expelled. The family money had sustained the school for five generations; the name was engraved on the hockey rink, the boathouse, and the memorial gate. What was at issue was the young man's continuing progress toward a mature appreciation of the art of good citizenship. The destruction of the library and the gym was obviously an accident; so was the loss of the five townspeople unfamiliar with the school's emergency procedures—regrettable, but not something that could have been avoided.

The headmaster didn't expect any mucking around in the swamp of vain regret (flowers had been sent, the lawyers paid, a stained-glass window donated to the village church), but at least it was conceivable that the boy might produce a few words of apology or remorse. Here he was on his best behavior in a stiff-backed chair (dark suit, blue tie, clean shoes), and why not say that he was sorry, didn't mean to hurt anybody, no sir, not what he had in mind at all.

Young Master Bush didn't condescend to tell so sad a story. Excuses were for scholarship students, not for the captain of the crew team. Certain that his conduct and deportment were noble, true, and right, he reminded the headmaster that the bonfire was the biggest one ever built in the history of the school; that he had wanted to show the proper spirit, to salute the undefeated football season, pay tribute to the parents and alumni proudly assembled on the lawn. It wasn't his fault that the wood was rotten and the intel less than perfect; nobody told him that fireworks sometimes have a way of getting crazily out of hand, and for the entire duration of the prerecorded hour he presented the headmaster not with an explanation, much less with an admission of possible error, but with a boastful flow of pious sentiment and a stalwart repetition of imaginary facts.

The President's performance was sufficiently awful to behold that it alarmed the gentry in the conservative as well as in the liberal press. The *New York Times* editorial page predictably found that "none of what we heard made much sense"; less predictable were the concerns voiced by the Republican loyalists on the reactionary right: Robert Novak ("This failure," "inadequate"), David Brooks ("Like most of us, President Bush doesn't have the facility for perfectly expressing his situation in conversation"), Peggy Noonan ("He seemed in some way disconnected from the event").

Even the bad reviews contrived to miss the point. Like the President's critics, the President's admirers make the mistake of assuming that he gives much of a damn about the intelligence product, about what does or doesn't happen in Iraq, about the success or failure of the steel tariff, the Environmental Protection Act, or the public schools. Although comforting, the assumption is impertinent. To the President's way of thinking, the only important story is the one about George W. Bush—what he feels and how he looks; Pontifex Maximus, the country's Celebrity in Chief, uninterested in history, lacking any frame of reference except the stage on which George W. Bush, the only actor in the play, must please George W. Bush, the only audience.

To Russert the President had said, "And the American people need to know they got a president who sees the world the way it is," which, of all the tales told to the NBC cameras, was by far the most fantastic. Mr. Bush sees the world the way he chooses to see it, preferably in a mirror. For the first time in the country's history, this year's federal budget has been illustrated with handsome four-color photographs, twenty-seven of them of the President—at the foot of the Washington Monument, in front of the American flag, blazing a trail through the Santa Monica mountains, teaching a small child to read the alphabet.

The narcissism is hereditary, not only within the Bush family but also within the American ruling and possessing classes that over the last fifty years have come to regard themselves as virtuous as well as rich, the masters and commanders of *Starship Earth*. The children of fortune learn to conceive the making of foreign policy as some sort of sporting event—a nation is slave or free, north or south, Christian or Muslim, "with us or against us." They believe themselves entitled to a view from the box seats or the deck of an aircraft carrier, from which vantage point, glory be to God and the science of naval architecture, the world presents itself as object, the United States as subject.

In war, Napoleon once said, the greatest sin is to make pictures, but the man who has inherited a great fortune does nothing else except make pictures. Unlike the poor man, who must study other people's motives and desires if he hopes to gain something from them, the rich man can afford to look only at what comforts or amuses him. He believes what he is told because he has no reason not to do so. What difference does it make? If everything is make-believe, then everything is as plausible as everything else. Jonas Savimbi can promise to go among the tribes and instruct them in the magic of constitutional self-government; the Shah of Iran can say that he means to make a democratic state among people who believe that they have won the blessing of Allah by burning to death 400 schoolchildren in a movie theater; the Iraqi exile Ahmad Chalabi, a known thief and notorious confidence man, can convince not only Dick Cheney but also Donald Rumsfeld and Paul Wolfowitz that the spirit of Adolf Hitler returned to earth in the body of Saddam Hussein.

To the extent that the business of the state becomes a matter of conscience, our presidents offer the expression of their private feelings as formulations of public policy, and foreign military intervention becomes a drama in the theater of the self. President Jimmy Carter promised to redeem the country, not to govern it; to the White House situation room he brought revelations instead of policies, envisioning a Palestinian state risen in the mist somewhere near the Sea of Galilee, informing a lobbyist from Boeing that he had decided against the deployment of a B-1 bomber because he had asked God about it, and God had told him that the bomber could do nothing but harm. During the week before the first President Bush decided to rescue Somalia from tyranny in the winter of 1992, the White House press corps found him moody and out-of-sorts, depressed about the loss of that year's election and complaining, only partly in jest, that he had nothing left to do except walk the dogs. His privy counselors recommended the Somalian adventure as a means of raising his spirits. In August 1998, three days after testifying about his penis before a Washington grand jury, President Clinton revenged himself against the stain on Monica Lewinsky's blue dress by sending cruise missiles into Afghanistan and the Sudan.

The American electorate doesn't require a presidential candidate to know where to look on the map for Romania or Zanzibar. His ignorance serves as proof of virtue. The man who would be president must present himself as an innocent and clean-limbed fellow, who knows nothing of ambition, murder, cowardice, or lust, and why would such a true American take the trouble to read the history of England, India, or Japan? He never has time to listen to the whole story or read through the long list of names that he doesn't know how to pronounce; he has planes to catch and meetings to attend, and his habit of inattention remands the making of the country's foreign policy to a cadre of Wall Street bankers and corporate executives who perform the service of family lawyers doing things in the heir's name but not in his sight.

Unfortunately for the hope of a democratic republic, the palace guard often drifts off into the same dream state that captivates the President. The documentary film *The Fog of War* illustrates this dismal tendency in an extended interview with Robert S. McNamara, the American secretary of defense during the Vietnam War years, 1961–68. The film was released in early January, but it so happened that I first saw it during the same week in which President Bush appeared in the Oval Office with Headmaster Russert, and I was struck by the parity of tone in the two performances. Like the President, McNamara doesn't hold himself accountable for his actions. Yes, mistakes were made and many people killed to no purpose (3.4 million Vietnamese and 150,000 Cambodians as well as 58,000 Americans), but killing is what wars are all about, regrettable but not to be avoided. Proud to have served his country ("some of the best years of our life . . . it was terrific!"), McNamara couldn't think of anything that he might have done differently, and in place of the wisdom that other men distill from suffering, he had collected a set of schoolboy maxims suitable for mounting on a wall of the Pentagon—"The human race needs to think more about killing," "In order to do good you may have to engage in evil,"

"Reason has limits," etc. He awarded himself the rank and title of "Military Commander," and mostly he talked about his feelings— not about what happened in Vietnam but about how it was up there on the great stage of world events with a wonderful guy by the name of Robert S. McNamara, as happily lost as President Bush in the fog of self.

April 2004

Buffalo Dances

There is no national science just as there is no national
multiplication table; what is national is no longer science.
—Anton Chekhov

*T*he reasons not to reelect President George W. Bush probably
number in the thousands, but I know of few as obvious as the
one put forward in a little-noticed report published last February
by the Union of Concerned Scientists under the unassuming title
"Scientific Integrity in Policymaking." Although signed by sixty of
the country's most accomplished scientists, among them twenty
Nobel laureates honored for their work in as many disciplines (mo-
lecular biology, chemistry, superconductivity, zoology, experimen-
tal particle physics, oceanography, etc.), the report didn't attract
the attention of news media preoccupied with the comings and goings of the winter primary campaigns. What the scientists had to
say gathered meaning not from new or sensational allegations but
from the collective witness of numerous individuals subjected to
various forms of censorship when called upon to provide scientific
data on matters of government policy. Invited to testify before a

federal agency or a congressional committee on topics as diverse as climate change, military intelligence, and the Missouri River, they found most of the questions already answered in compliance with the Bush Administration's prepaid and prerecorded political agenda. Time and again in the thirty-eight-page report the respondents mention the stone-faced displays of willed ignorance, and their plain statements of fact add to the sum of an indictment more telling than Senator John Kerry's best efforts at emotional diatribe and quarrelsome polemic:

> It's hard not to think that our findings don't match up with what they want to hear, they are putting a new team on the job who will give them what they want.
>
> . . . restrictions on what government scientists can say or write about "sensitive" topics.
>
> . . . well-established pattern of suppression and distortion of scientific findings . . .
>
> No fewer than twelve paragraphs lifted, sometimes verbatim, from a legal document prepared by industry lawyers.
>
> Information suggesting a link between abortion and breast cancer was posted on the National Cancer Institute website despite objections from CDC staff.

The report speaks to the presence of something new under the American sun. Not the art of shading the statistics to accommodate a specific commercial or political interest—a familiar practice, as well established as the hiring of competitive teams of experts—but the systematic substitution of ideological certainty for reasonable doubt across the entire spectrum of issues bearing on the public health and welfare. In years past the testimony often and predictably brought forth sharp differences of interpretation of almost any fact that anybody cared to name, whether in regard to the life span of a bacteria or the trajectory of a cruise missile, but nearly everybody in the room could be relied upon to employ the instruments of logic and deductive reasoning and at least to acknowledge more or less the same rules of evidence.

The Bush Administration apparently has declared the expectation inoperative, and with it the principle on which the country is founded. The American democracy depends for its existence on the force of reason and the uses of experiment, and if I read correctly the report from the Union of Concerned Scientists, the signatories find fault with the Republican college of augurs in Washington not for a single error (or even a multiple choice of errors) but for its rejection of the scientific method in favor of the conviction that if the science doesn't prove what it's been told to prove, then the science has been tampered with by Satan or the Democratic Party.

The disdain for disloyal or unpatriotic fact defines the Bush Administration's approach not only to questions likely to embarrass the oil, weapons, and insurance industries but also to those that might interfere with its fanciful conceptions of war and money. The invasion of Iraq went forward with the blessing of counterfeit evidence (about the weapons of mass destruction and Saddam Hussein's alliance with Al Qaeda), as did the passage of the Medicare prescription drug bill (the known cost of $500 billion reduced to a more convenient $400 billion); Christine Todd Whitman resigned her cabinet appointment to the Environmental Protection Agency because she couldn't stomach the White House's instructions to deny the ill effects of carbon dioxide in the atmosphere; General Eric Shinseki, the Army's chief of staff, retired after telling Congress that the number of American troops required to occupy Iraq would come closer to "several hundred thousand" than to the

100,000 confidently anticipated by Secretary of Defense Donald Rumsfeld; Paul O'Neill was dismissed as the secretary of the treasury because he refused to say that money grows on trees. To read the stories in the papers is to remember that in Nazi Germany, Einstein's theory of relativity was condemned as a Jewish corruption of Isaac Newton's "organic" laws of nature, and that in Stalinist Russia, biologists were first tortured and then executed for failing to admit that the science of genetics was a "reactionary" and "bourgeois" lie.

The Union of Concerned Scientists concludes its report with the expression of a hope that the government's divorce from reality is a curable disease subject to congressional diagnosis and legislative remedy. I wish I could share the supposition. Although I can appreciate the scale of the collective alarm, I think it late in arriving and wrongly placed. The reality to which the signatories refer went out of fashion some years ago, and it is they, not the Bush Administration, who find themselves at odds with the forms of magical thinking that now constitute the society's surest proofs of wisdom. Why else does America declare a war on terrorism that resembles a geopolitical video game adapted from the script of *Lord of the Rings*? How is it possible to elect Arnold Schwarzenegger governor of California except in the belief that he will bring to Sacramento the secret stone of power that Conan the Barbarian rescued from the sorceress in the castle of Shan?

The postmodern sensibility is a product of the electronic media, which lend themselves more readily to the traffic in dreams and incantations than to the distributions of coherent argument. As the habits of mind beholden to the rule of images come to replace the systems of thought derived from the meaning of words, the constant viewer learns to eliminate the association of cause with effect.

Within the walls of silicon and glass that enclose the artificial king-dom known to the late Marshall McLuhan as "the pool of Narcis-sus," the time is always now—if not on channels 1 through 524 in New York or Los Angeles, then somewhere among the blogs floating in the sea of cyberspace; if not on *Oprah* or the *Drudge Report,* then at votenader.org or at the award-winning adult site www.whitehouse .com. In the enchanted garden of the eternal present John F. Kennedy is still the king in Camelot and Jane Austen is forever rid-ing in a carriage on the road to Bath. Weightless and without con-sequence, the images drifting across the mirrors of the self appear and disappear in no context other than their own, demanding noth-ing of the audience except the duty of ritual observance. Because the camera sees but cannot think, it doesn't matter who sings the undying songs of love, or whether the twenty-four-hour circus pa-rade goes nowhere except around in circles. Nothing necessarily follows from anything else; what is important is the surge and vol-ume of emotion, not its object or its subject, and the accelerated data streams of the virtual future carry the friends of Rush Lim-baugh backward into the firelight flickering in the caves of the pagan and prehistoric past. Narrative dissolves into montage and knowledge becomes a matter of instantly recognizing the iconogra-phy (Osama's beard, the Nike swoosh, Ralph Lauren's polo player, Howard Dean's upraised fist); history reverts to myth, and politics collapse into the staging of pageants sometimes accompanied by a fall of brightly colored balloons.

Similar in character to the reality television shows constructed along the lines of *Survivor* and *The Apprentice,* the election cam-paigns borrow their heroics from the medieval chroniclers who told of Christian knights sent in search of dragons, required to recover bits and pieces of the true cross and to wander for many days and nights in a wilderness of elves. In the early years of the twenty-first century, in a country that boasts of its prowess with laser beams, the man who would be president must endure the trials by klieg light and wander for many days and nights in a labyrinth of Holiday Inns, fending off the banquet food and hurling press releases at the other candidate from Yale, answering, in twenty words or less, questions that can't be answered in four or twenty thousand. The presidency undoubtedly constitutes a fearful test of a man's capacities, but his capacities for what? Even if the electorate understood or cared about the tedious business of government, how does it choose between the rivals for its fealty and esteem? The one attribute that can be known and seen comes to stand for all the other attributes that remain invisible, and so the test becomes one of finding out who can survive the stupidity and pitiless indifference of the television cameras.

Of our presidents we make celebrities—a safer form of constitutional divinity than the one embodied in the name of Caesar— and with gifts of adulation and applause the media appoint them to the office of a totem pole. To say that Mr. Bush is an incompetent president is like saying that Tom Cruise can't act or that Britney Spears can't sing. The observation might be true, but it ignores the point that celebrity is a commodity meant to be sold at the supermarket checkout counter with the cosmetics and the canned soup. What was once a subject has become an object, no longer capable of error or human speech, imparting a sense of stability and calm to a world in which the science or the military intelligence refuses to prove what it's been told to prove. The headlines bring word of death in Iraq and terrorism in Spain, famine in Somalia and banditry in Washington, but on the smooth and reassuring surface of a magazine cover or movie screen, the golden masks of divine celebrity—George W. and Nicole, Julia and the Donald—bestow upon the faithful the smiles of infinite bliss.

*   *   *

The magical forms of thinking inhabit so many different quarters of the society that it's hard to read the morning paper, much less watch the evening news broadcasts, without coming across another instance of the broad retreat into the forests of superstition. President Bush goes before a crowd of factory workers in Ohio, many of them recently unemployed, and while posed in front of a bright, blue banner announcing the miracle of JOBS AND GROWTH, he says, with a boyish and cheerful grin, that he intends to tear down the nation's remaining trade barriers and make permanent the tax cut for the very rich. The attending press corps doesn't comment on either the witlessness or the inappropriateness of the remark because it is recognized as the casting of a sacred spell. All present understand that Mr. Bush might as well be waving Pharaoh's wand or performing the Arapaho buffalo dance. Among people who worship the objects of their own invention, whether in the form of a satellite phone or as the chrome-plated image of a president, ritual becomes a form of applied knowledge.

Thus the House of Representatives when it considers legislation designating cheeseburgers as enemies of the state, or Michael R. Bloomberg, the mayor of New York, when he issues an edict requiring every third-grade student in the city's public-school system to pass a standardized test (engineered by academic soothsayers and issued by a computer) before being allowed to advance to the fourth grade. It doesn't matter that the test provides no accurate measure of the child's intelligence and no fair assessment of the child's needs or circumstances. The test looks for a sign from Heaven, the presumption similar to that of a twelfth-century Catholic bishop seeking to discover a sinner's guilt or innocence by forcing him to walk on hot coals. The season's most successful movie ($295 million at the box office in the first four weeks of its release) is Mel Gibson's *The Passion of the Christ*, which devotes 100 of its 126 minutes to the bloody slaughter of a sacrificial animal that would have gladdened the mob in the Roman Colosseum, and Martha Stewart suffers the scourging of a criminal trial for having failed in her duties as a vestal virgin.

Although I don't doubt that a society in which fewer and fewer people know how to think is probably easier to manage than one in which too many people ask too many questions for which they don't already know the answers, the flight into the self-referential landscapes of wish and dream doesn't hold out much promise for the American future. It speaks instead to the exhaustion of the spirit and intent that framed the Constitution.

The inventors of the country's liberties recognized themselves as scientists, makers of maps and collectors of beetles, who pursued their studies in as many spheres of reference as could be crowded into a Philadelphia library company or a Boston philosophical society. Together with Benjamin Franklin they delighted in the joys of discovery and understood the American democracy as an ongoing experiment with the volatile substance of freedom.

Both of this year's presidential candidates acknowledge the country's recent loss of 3 million manufacturing jobs, but neither of them mentions the more serious wound, which is the depletion of the national reserves of political intelligence. To the extent that we take comfort in the illusion that the future can be bought instead of earned, we join Secretary of Defense Rumsfeld in the magical explanation for the non-existence of hideous weapons in Iraq ("The absence of evidence is not evidence of absence") or Attorney General John Ashcroft in the belief that in America "there is no king but Jesus."

The phrase "national security" undoubtedly will make numerous appearances in the campaign speeches between now and the November election, and if the ritual holds true to form it will add to the country's inventories of fear instead of increasing its store of courage. To define the national security as a wonder of aircraft

carriers or a marvel of surveillance cameras is to mistake the lesser
for the greater instruments of American power, to miss the point,
made by the signatories to both the Constitution and the report
from the Union of Concerned Scientists, that the republic's best
and only chance for survival rests on its freedom of thought and
force of mind.

May 2004

That's Why the Lady
Is a Champ

This is the excellent foppery of the world, that, when we are sick in fortune,—
often the surfeit of our own behaviour,—we make guilty of our disasters the sun,
the moon, and the stars: as if we were villains by necessity; fools by heavenly
compulsion; knaves, thieves, and treachers, by spherical predominance. . . .
—William Shakespeare, *King Lear*

Not having read through all the testimony presented to the 9/11 commission over the last fourteen months, I can't say with certainty that no witness blamed the loss of the World Trade Center on an eclipse of the sun or an untimely rising of the September moon. Attorney General John Ashcroft may have attributed the calamity to Satan as well as to the Clinton Administration, and it's conceivable that George Tenet named as villain an irregularity in the orbit of Mars; the newspapers didn't make available the complete transcript of their remarks, and one or another of the redacted paragraphs possibly mentions a spherical predominance known only to the astrologers at the Justice Department and the CIA.

But if the excuses unincorporated in the public record must be given up for lost, those brought forward into the light of the television cameras offered enough instruction (how to frame the unintelligible answer, when to introduce the meaningless phrase, with which finger to appoint the blame) that they can be understood not only as proofs of the Bush Administration's divorce from reality but also as a comprehensive exhibit of what has become the state of the art of self-exoneration.

Misjudgment in high places can be accepted as a constant; so can the habits of mind that favor criminal incompetence and the pretensions to imperial splendor. I don't find it surprising that when the United States was attacked by Saudi Arabian jihadists we responded by attacking a secular regime in Iraq. Germany in 1914 declared war on Russia and invaded France. All governments enchanted by the story of their own magnificence fall afoul of the same stupidities—the waste of money, the misreading of foreign intelligence reports, the breakdown of interior lines of communication, the inability to see further than three feet into the future. The variable is the character and quality of the excuses.

Fifty years ago in the United States the appearance of the truth was still a valuable commodity, which obliged the witnesses lying under oath to practice the techniques associated with the realist schools of nineteenth-century landscape and portrait painting. Public servants who appeared before Congress in the 1950s either had to honor the conventions then in place or withdraw into the silence protected by the Fifth Amendment. It wasn't simply a matter of conduct and deportment; the members of the committee were old-fashioned people, not yet postmodern, and apt to take a genuine interest in the literary distinctions between fiction and nonfiction. Presidents John F. Kennedy and Lyndon Johnson accepted responsibility for their mistakes in Cuba and Vietnam, and President Richard Nixon was impeached for his unwillingness to produce even a facsimile of the truth.

The style had changed by the time the Washington theater company staged its 1987 production of the Iran-Contra hearings. Ronald Reagan was president (a man who was himself hard-pressed to know what was true and what was not), and it was no longer necessary to work up a plausible alibi. It was enough to lower one's voice to an agreeable murmur and say, with a gracious and accommodating smile, "I can't recall." Colin Powell, then senior military assistant to Secretary of Defense Caspar Weinberger, exercised the option 56 times when he refused to tell Congress why the United States had sold missiles to despotic mullahs in Iran in return for money with which to fund, secretly and illegally, a fascist junta in Nicaragua. President Bill Clinton in the 1990s experimented with subtle refinements of language—"it depends upon what the meaning of the word 'is' means" or, "Well, again, it depends on how you define 'alone'"—but he was too much of an avant-garde figure to establish a reliable school of rhetoric. Most people didn't possess either his charm or his talent for improvising lectures on the importance of the universe that diverted unwelcome questions into a fog of boredom.

The senior managers of the Bush Administration came to Washington knowing how to manufacture disinformation in commercial quantity; appreciative of the uses of advanced technology and familiar with the idioms of opaque abstraction ("contingency planning process," "polycentric multi-polar paradigm," etc.) acquired during their years of experience in the country's corporate boardrooms and reactionary policy institutes. Secretary of Defense Donald Rumsfeld speaks with the authority of a man capable of falsifying both a balance sheet and the history of Egypt. Well known for teaching his deputy, Paul Wolfowitz, how to address a news conference ("Begin with an illogical premise and proceed perfectly logically to an illogical conclusion"), Rumsfeld is also fondly remembered for responding to an inquiry about the missing weapons of mass destruction in Iraq with an utterance that the Pentagon spokespersons could as easily have attributed to the oracle at Delphi: "The absence of evidence is not evidence of absence."

Testifying before the 9/11 commission on March 23, Rumsfeld deployed a deft non sequitur to avoid answering a question as to whether American Air Force pilots were under orders on September

11 to shoot down civilian airplanes presumably under the control of terrorists. "I do not know," said Rumsfeld, "what they thought." Two weeks later at a Pentagon press briefing on the bad news then incoming from Iraq (forty American soldiers killed in sudden uprisings everywhere in the country, 1,500 Marines fighting in the streets of Fallujah, the charred and mutilated bodies of two American civilians hung from a bridge), Rumsfeld ascribed the misfortune to the hand of fate, beyond the comprehension of merely mortal men:

"We're trying to explain how things are going," he said, "and they are going as they are going. . . . Some things are going well and some things obviously are not going well. You're going to have good days and bad days."

The secretary's performance was admired both by the board of editors at the *Wall Street Journal* and the militant resistance in Baghdad, but it was no match for the genius of Condoleezza Rice, the national security adviser, who appeared before the 9/11 commission on April 8 to answer questions about the government's laissez-faire attitude toward terrorism during the spring and summer of 2001. The circumstances were by no means favorable to the off-loading of accountability onto a wayward planet or a distant star. The front-page newspaper headlines that morning were continuing to bring word of dead Americans in the deserts south and west of Baghdad, and for three weeks Washington had been consumed with gossip about the government's failure to heed repeated warnings of an imminent attack on the United States by the agents of Al Qaeda. Richard Clarke, a former counter-terrorism operative within the Bush Administration, was making the rounds of the Sunday talk-show circuit to say, loudly and convincingly, that the White House had done "a terrible job" of protecting the country against the rapidly multiplying host of its enemies. The available facts tended to support the judgment.

Not the most congenial of atmospheres in which to float the balloons of innocence. Dr. Rice managed to do so by postulating a world in which actions bear no relation to their consequences. Speaking only in the dialect of government acronym and careful to stay within the persona of the brightest girl on campus, Dr. Rice confined her testimony to a world made entirely of paper. Her office had filled out all the necessary forms, followed all the correct procedures, composed the appropriate number of power points. Meetings had been held, documents distributed, FBI field offices "tasked" to increase surveillance, the CIA "advised" to examine its conscience and search its data banks. Had the national security adviser known that death was destined to come to Manhattan from a clear blue sky, she would have moved heaven and earth to send it somewhere else, but she didn't know, and the moving of heaven and earth was a topic that President Bush discussed only with God. Dr. Rice was responsible for moving paper; she had moved it in all the ways it could be moved, and if some of it got lost or went to the wrong address, no human being was at fault. The fault was in the "structure" of the country's systems of foreign and domestic intelligence that on arriving in Washington she had found to be as dysfunctional as the stones on Easter Island. A legion of engineers would need a decade to teach all the computers how to talk to one another; the Bush Administration had at its disposal only 233 days, and in the meantime it was the "structure," also the "process," that made us guilty of our disasters.

Careful to draw the distinction between "strategic" and "tactical" thinking, Dr. Rice explained that it was no good pursuing terrorists in an ad hoc manner, one or two at a time as they happened to show up at random in a German night club, on an Indonesian beach, or attending flight schools in Phoenix and Minneapolis. Overly eager crisis managers like Richard Clarke might want to seize terrorists whenever and wherever they could be found, but this was a mistake, entirely the wrong approach, exposed to gunfire and subject to confusion. Terrorists must first be captured within a policy, a global as well as a regional policy, both supported with maps, briefing papers, and a diplomatic impact statement.

"America's Al Qaeda policy wasn't working because our Afghanistan policy wasn't working," Dr. Rice said, "and our Afghanistan policy wasn't working because our Pakistan policy wasn't working."

The Republican commissioners were quick to acknowledge the rightness of the argument (of course, Pakistan, the source of all our sorrows), but some of the Democrats in the room were still curious as to why the President had paid so little attention to a CIA briefing paper presented to him in Texas on August 6 under the rubric "Bin Ladin Determined to Strike Inside the United States." The memorandum referred to the World Trade Center, to the FBI conducting seventy field investigations of Al Qaeda cells within the United States, also to "patterns of suspicious activity in the United States consistent with preparations for hijackings . . ."

Confronted with the most potentially damaging question put by the commission, Dr. Rice found in her answer the point at which her testimony ascended into the heavens of the sublime. The fault was not in man or beast or the constellation of Orion; the fault, my lords, was in the threat itself. It wasn't sent by courier or distributed in triplicate, "was not a warning," because "it was not specific as to time, nor place, nor manner of attack."

Affronted by the presumption of a threat so vaguely worded and improperly presented, Dr. Rice dismissed it with an air of exasperated impatience. Who did these people think they were? What kind of third-rate terrorist operation forgets to enclose a return envelope? Perhaps the commissioners were unaware of the complexity contained within the national security adviser's office—the number of top-secret bulletins arriving from China and the Balkans as well as from Africa and the Middle East, the difficulty of sorting out the strange accents on the telephone intercepts, the brute weight of paper from which somebody had to remove the stains of human error. Under the circumstances, who had time to decode, much less take seriously, the angry muttering of every Arab in Djibouti?

For three hours in front of the commission Dr. Rice never once abandoned her position within the capsule of virtual reality, her virtuoso performance acknowledged by the next morning's headline in the *New York Post* ("The Lady Is a Champ") and by President Bush, who telephoned her from his pickup truck in Texas to offer his congratulations. Not only had she mollified the administration's critics; she had proved the worth of the new school of sophistry that borrows its aesthetic from surrealist painting and Dadaist poetry. On the ground in Iraq the Bush Administration's imbecile foreign policy had been reduced to a shambles of burning automobile tires—the several Iraqi religious factions hating America more than they hated one another, the Coalition Provisional Authority at a loss to know how or to whom it was bound to transfer sovereignty on June 30, no foreseeable end to the killing in the streets—but in the hall of mirrors that surrounds Dr. Rice the American occupation glowed with the smile of victory because it has removed the "sources of violence and fear and instability in the world's most dangerous region."

Having the courage to tell it like it isn't is why the lady is a champ. Simply by translating the story into the language of blameless abstraction she not only removed the fear and violence from Iraq but also lifted from the Bush Administration the burden of its mistakes. Everything was in order in the world of paper (was then, is now, forever will be); it was the events that were at fault, the incorrectly processed events and the treacherous office equipment.

*   *   *

The high-end Washington journalists say that President Bush gazes upon the pedantry of his national security adviser with the reverence that Alexander of Macedon once showed to Aristotle, and on the evidence of her testimony before the 9/11 commission I can understand why he does his best to imitate her rhetorical style. The technique is as easy to learn as the vocabulary of corporate advertising and, with the proviso that one must keep a straight face, a form of address that lends itself to the earnest recitation of heartfelt nonsense.

At his press conference on April 13 the President not only borrowed a number of tropes from Dr. Rice's tour de force on Capitol Hill (had we but known, "We would have moved heaven and earth"; "we hadn't got our relationship right with Pakistan yet," etc.) but also demonstrated his steadily improving skill at the moral vanishing act. Mr. Bush prefers the religious to the secular forms of escape into unreality, and so instead of appealing to the spherical predominances found in the realm of policy, he transferred the reason for the American presence in Iraq to a heavenly compulsion: "Freedom is the Almighty's gift to every man and woman in this world. And as the greatest power on the face of the earth, we have an obligation to help the spread of freedom." Also the obligation to remember that what is real is the theory of war that appears on the maps and the computer screen, not the experience of war that blots the flow charts with the smudges of human suffering, mutilation, and death.

June 2004

Chasing the Pot

Though the object of being a Great Power is to be able to fight a Great War,
the only way of remaining a Great Power is not to fight one.
—A.J.P. Taylor

*P*oker players who win more often than they lose obey a rule of thumb expressed in the phrase "never chase the pot." Were the United States to apply the same policy to the cards it has been dealt in Iraq we would fold the hand sooner instead of later—without conditions or complaint, accepting the loss as a fact beyond the hope of rescue by Commander James Bond or the ace of hearts.

President George Bush apparently doesn't know the game (doesn't know it or believes himself too rich to care what any of the numbers mean), and so in the news from Washington and Baghdad these days we see the squandering of the country's fortune (its wealth, the lives of its young men and women, its character and good name) on the vanity of a feckless commander in chief who holds the equivalent of five low and unmatched cards—a bankrupt theory of world domination, a collection of lurid snapshots from the Abu Ghraib prison, a botched military occupation of the Mesopotamian desert, a delusionary secretary of defense, few allies in western Europe and none in the kingdoms of Islam. Undeterred by circumstance, well pleased with his persona as the last, best hope of mankind, the President smiles his spendthrift and self-congratulating smile and bets another Marine division on the chance that it will save Mel Gibson's Jesus from a mob of bearded terrorists in Najaf.

I can understand why some people might find the performance terrifying, also why some other people might see it as darkly comic, but what I don't understand is why anybody continues to think that the man knows what he's doing. Presumably they're unacquainted with the lessons of the poker table; maybe they don't know that the President imagines himself in a game with John Wayne, Omar Sharif, and the Devil. Important personages in the news media, sources well informed and highly placed, acknowledge the mess that the noble heir has made of the American gamble in Iraq, but when I suggest that the President would do well to heed the advice of the historian A.J.P. Taylor, the tribunes on the jingo right accuse me of cowardice or treason (not a true American, no friend of our soldiers in the field); representatives of the conscience-stricken left draw my attention to the geopolitical reality of the international oil price and Woodrow Wilson's high-minded notion of making the world safe for democracy. But no matter what the provenience of the correction or the rebuke, all present in the chorus of responsible opinion (Senator John Kerry as well as President Bush) offer sentiments identical to the ones that for twelve years bankrolled the American losses in Vietnam—the United States must "stay the course," discharge its "moral responsibility," protect the Iraqi people from the scourge of civil war, maintain its "credibility" as the all-powerful wonder of the world. The sales pitch is as disingenuous now as it was in 1968:

AMERICA MUST FINISH THE JOB

What job? Instead of going to Iraq with plans for a military occupation, we went with the script for a Hollywood western, and we have done as much as we know how to do—captured the bad guy, discovered that he didn't possess weapons of mass destruction, expended large quantities of ammunition, reduced to rubble a substantial weight of antiquated architecture, killed or maimed 4,000 American troops as well as an unlisted number of Iraqi civilians. Beginning and end of story, consistent with the plotline of *High Plains Drifter* and similar in both disposition and result to the American expeditions to Cuba and the Philippines at the turn of the twentieth century, to Haiti in 1915 and 1994, to Vietnam in 1962–75.

Certainly we can do more of the same, but as to the construction of a democratic theme park on the banks of the Euphrates, we have neither a liking nor a talent for the enterprise. Of the books recently leading the *New York Times* bestseller list, at least six (among them those by Richard Clarke, Ron Suskind, John Dean, and Bob Woodward) speak to the Bush Administration's failure to comprehend, much less anticipate or address, the task described in the Pentagon's promotional brochures. The Washington aides-de-camp arranging the maps on the walls of the White House Situation Room tend to overlook the fact that the Americans are an authentically civilian people, devoid of an exalted theory of the state that might allow us to govern subject races with a firm hand and a quiet conscience. The imperial project serves the interest of the propertied classes, but the work must be performed by the laboring classes, and it is never easy to harness the energy of the latter to the enthusiasms of the former. At about the same time that the Abu

Ghraib prison photographs were making the rounds of the Internet, the apostles of freedom-loving blitzkrieg were acknowledging the need for another fifteen infantry divisions to stand watch for another four or five years over the investments of American capital in the oil fields of Basra and Kirkuk. The Pentagon doesn't have the troops, and if the opinion polls can be believed (popular support for the liberation of Iraq falling from 53 percent to 48 percent in the month of May) the American public isn't willing to spend the money or the time. Who then will finish the job? None other than the Iraqi people, currently identified (as were the rebels in the American revolution) as terrorists, detainees, foreign agents, enemy combatants.

MORAL RESPONSIBILITY

The two words sound nicely together when presented to a television camera or to a conference sponsored by the American Enterprise Institute, but where is the meaning in either the adjective or the noun? How moral? Responsibility to what or to whom?

I know of no war described by its active participants as moral. Men go off to war for as many reasons as can be dressed up with a flag or a name for God; the horror of the battlefield translates the fine language into the savage instinct for survival, the military band music into the sound of dying animals. Consult the testimony of the witnesses to the killing at Verdun, Stalingrad, or Omaha Beach, and the voices of fast-receding conscience find nothing moral in the mass production of their own collective murder.

The American diplomatists of fifty years ago at least had the good grace not to mince words. George Kennan in the winter of 1948 circulated a memorandum to his associates at the State Department in which he made no attempt to conceal the motive of straightforward plunder behind the screens of Christian charity:

We have about 50% of the world's wealth, but only 6.3% of its population. . . . In this situation, we cannot fail to be the object of envy and resentment. Our real task in the coming period is to devise a pattern of relationships which will permit us to maintain this position of disparity without positive detriment to our national security. To do so, we will have to dispense with all sentimentality and day-dreaming.

The argument presupposed an American realpolitik strongmind-edly turned away from what Kennan regarded as "unreal objec-tives such as human rights, the raising of the living standard and democratization."

Responsibility, the second word in the stock phrase, is as non-sensical as the first. The invasion of Iraq was mounted on the false premise that secular democracy (that happy, blessed state) could be forced at gunpoint upon Muslim nationalists faithful to the teachings of the Koran and more familiar with the punish-ment inflicted upon them by ten years of American economic sanctions than with the sayings of Thomas Jefferson. To excuse the subsequent fiasco of the military occupation, the President asks the American people for the willing suspension of their dis-belief. Although characteristic of an administration that defines the acceptance of responsibility as a fool's errand and the admis-sion of error as a sign of weakness, the stratagem comes up against a difficulty on which a younger and wiser John Kerry re-marked soon after his return from the war in Vietnam, a war in which he'd served with distinction as a naval lieutenant but had come to regard as both irresponsible and immoral. Testifying be-fore the Senate Foreign Relations Committee in April 1971, he said, "How do you ask a man to be the last man to die for a mis-take?" The Bush Administration answers the question with the directive, "Tell the man a lie."

AMERICA'S CREDIBILITY AT STAKE

Not America's credibility, the credibility of the Bush Administration. The phrase "American foreign policy" currently stands as a synonym for cynicism and deceit, the credit rating of a White House press release so sharply discounted nearly everywhere in the world among people unaffiliated with the weapons industry, a law-enforcement agency, or the Republican Party that in Israel the newspapers defend the cruelties inflicted on the Palestinians in Gaza by saying, "This isn't America; the government did not invent intelligence material nor exaggerate the description of the threat to justify their attack."

The Bush Administration staged the violent overthrow of Saddam Hussein to prove that America's colossal military power established its right to rule the world from the gun platforms of virtual omnipotence. During the first weeks of the invasion the staff officers at the White House congratulated one another on "the demonstration effect" of their high-tech gladiatorial show in the cradle of civilization, certain that the fireworks display would so shock and amaze troublesome regimes elsewhere in the Middle East that no Arab in his or her right mind would chance the risk of overt or covert defiance. Contrary to the expectations of the studio executives in Washington, the events of the last year have taught a different object lesson, demonstrating the limits of American power and suggesting that the Bush Administration's imperialist policy amounts to little else except another name for terrorism—precision-guided and electronically accessorized but otherwise similar in its objectives to the action-movie sequence that destroyed Manhattan's World Trade Center. The apostles of "the new information order" have been making the point for twenty years. The terrorist who blows up a train in Madrid enlists the complicity of CNN, and within an hour of committing the atrocity, he holds as hostage the rage and despair of an audience large enough to wreck a government. So also the scenes of street fighting in Fallujah and the exhibition of photographs from the Abu Ghraib prison, seen within the hour by America's prospective enemies everywhere in the Islamic world. The war on terror is a war of images, the firepower of the world's television cameras striking an asymmetric balance against the weapons of mass destruction in the Pentagon's arsenal of fear.

Like the rescue of Vietnam from the evil castle of Soviet Communism, the rescue of Iraq from the caves of Arab jihad has borne out the law of unintended effects. Meant to astonish the world with the virtues of democracy, the expedition to Indochina taught a generation of American citizens to think of their own government as an oriental despotism. In Iraq we meant to render futile both the theory and the practice of terrorism; what we have done instead is to endow it with diplomatic credentials, making credible the policies of blind assassination.

FAILURE IS NOT AN OPTION

Define the American purpose in Iraq as the transformation of the Arab Middle East into a democratic real estate development, or the seizing of what President Bush fondly describes as "an historic opportunity to change the world," and failure is the only option. What is at issue is the degree of failure, and whether the United States can earn a measure of respect (from an increasingly large body of its own citizens as well as from the Iraqi people) by departing, preferably before the end of the year, without attempting to secure the perimeters of a puppet government or a client state. Every day that American troops continue to kill or be killed (for whatever reason on no matter whose orders) adds to the sum of anger and resentment certain to make more difficult the country's struggle with its own tribal hatreds, nationalist politics, religious zeal. We cannot know if our withdrawal will incite civil war, or, if such a war occurs, whether it will lead to a worse or more far reaching result than the one assured by our extended military presence. Our Washington geopoliticians wrongly forecast a massive bloodletting in Vietnam after our escape from the roof of the embassy in Saigon. Probably we won't like the government that the Iraqis choose for themselves— whether a secular state in control of its own oil reserves or an Islamic theocracy unfriendly to Israel—but if we mean what we say about democratic principle and free elections, on what ground do we prevent them from choosing it? Under the auspices of the United Nations, we can provide money and medicine, also roads, sewers, electrification, and copies of *The Federalist Papers*, but not a constitution similar to the one that we imposed on Cuba in 1901, which reserved our right to "intervene for the preservation of Cuban independence."

In the meantime we can hope that John Kerry, the presidential candidate, remembers the question asked by John Kerry, the naval lieutenant. For the time being he appears to have forgotten what he once knew about lost wars and dishonest advertising slogans. His campaign speeches echo the sentiments of his opponent (job to do, moral responsibility, more American troops backing the currency of Iraqi independence); whether offered in good faith or deemed politically expedient, the dead language condemns him to the prison of the past, to the belief that America's "national security" depends on the weight of its military power. The Bush Administration's gamble in Iraq has proved the error of the hypothesis. The country is less secure now than it was a year ago, the multiplication of our enemies outpacing our production of lying press releases and our manufacture of high-performance artillery shells.

The summer election season presents Kerry with the chance to find his way out of the hall of old mirrors, possibly to discover an exit strategy in the idea that our national security stems from the character and the intelligence of the American people—i.e., from the investment in education, health care, our own infrastructure and environment, not from the chasing of the country's fortune into the mouth of omnivorous and never-ending war.

July 2004

Crowd Control

A function of free speech under our system of government is to invite dispute. It may indeed best serve its high purposes when it induces a condition of unrest, creates dissatisfaction with conditions as they are, or even stirs people to anger.
—Justice William O. Douglas

Not mincing the words in its title, "The Assault on Free Speech, Public Assembly, and Dissent," the National Lawyers Guild last August published a report describing the coercive methods—rapidly multiplying and increasingly sophisticated—employed by the country's law-enforcement agencies to quiet the noise and, by so doing, sap the strengths of the American democracy. The evidence gathered from witnesses in Boston, Albuquerque, New York, Portland, Washington, and Miami speaks to the pathology of a government so frightened of its own citizens that it regards them as probable enemies. Whether elected to Congress or appointed by the news media, invested with the powers of a trial judge or a police sergeant, the wisdoms in office find the practice of democratic self-government obnoxious, tiresome, inefficient, loud, disrespectful, and unsafe. Let too many freedoms roam around loose in the streets, and who knows if or when they might ring off key, or whether somebody might turn up with an improvised explosive device or a copy of the Constitution.

The report takes as its cases in point sixteen large-scale street demonstrations organized over the last five years to protest—i.e., to question, oppose, disagree or quarrel with—a judgment, policy, decree, or act of government. Most of the demonstrations voiced objection to the invasion of Iraq—those in Washington, D.C. (September 2002), New York City (February 2003), Albuquerque, New Mexico (March 2003), again in Washington (April 2003); several others sought more enlightened debate on the topics of the environment and foreign trade, chief among them those staged at the Seattle meeting of the World Trade Organization in November 1999, in Washington at the meetings of the World Bank and International Monetary Fund in April 2000 and September 2002, at the New York meeting of the World Economic Forum in February 2002, and at the Miami conference on the Free Trade Area of the Americas in November 2003. No matter what the venue or the subject in dispute, the evidence invariably shows the civil authorities adapting to a domestic purpose the same military rules of engagement—preemptive strike, forward deterrence, anticipatory self-defense—that distinguish the Bush Administration's never-ending war on foreign terrorists. No breaking through the rope-lines of consensus, every citizen assumed guilty until proven innocent, unlicensed forms of speech deemed unpatriotic and disloyal, the protections of the First, Fourth, Fifth, Sixth, and Fourteenth amendments declared inoperative or canceled because of rain.

Careful to avoid inflammatory rhetoric, the report contents itself with plain statements of fact, presenting a catalogue of the new and improved means of crowd control that classify the freedoms of expression as a criminal offense. To the following brief summary of the text I've appended, in parentheses, the number of the Constitutional Amendment that each tactic vitiates or ignores:

POLICE INFILTRATION AND SURVEILLANCE (I, IV, AND XIV):

As amended in May 2002, the Justice Department's Guidelines on General Crimes, Racketeering Enterprise, and Terrorism Enterprise Investigations permits the FBI to engage in domestic espionage. At or about the same time the Department of Homeland Security issued an all-points bulletin advising the nation's law-enforcement agencies to keep a sharp watch for any citizens who might have "expressed dislike of attitudes and decisions of the U.S. government." Both sets of instruction encouraged more frequent investigations of known or suspected activists, if for no other reason than to furnish such people with the embarrassment of an arrest record and to induce among them a paranoid trembling of mind sufficient to render them stupid as well as futile.

The intimidations take multiple forms—reminders that any withholding of information can be punished as a crime, an impromptu search of an attic or a garage, casual mention of an email sent by the person under investigation to a friend in Paris or Islamabad. During the months prior to last summer's political conventions, FBI counterterrorism agents asked scores of people everywhere in the country whether they had any plans to travel to New York or Boston, and, if so, whether they intended to bring guns and/or seditious thoughts.

When conducting less forthright inspections of the citizenry, undercover FBI agents and police spies join protest movements to collect names and addresses, whenever possible to take photographs and make tape recordings; while participating in a demonstration or a march they sometimes act as *agents provocateurs*, shouting obscenities, breaking windows, causing enough of a disturbance to provoke a violent response from the uniformed police.

RUSH TACTICS (I, IV, AND XIV):

When police officers assault otherwise peaceable demonstrators without reasonable suspicion or probable cause, for no reason other than to instill the habit of obedience and preach the lesson of fear. The procedure frequently warrants the technique of riding into a defenseless crowd on horses or motorcycles.

Helicopters hovered over the heads of a crowd gathered nearby the building in which the Democratic National Convention was being held in Los Angeles in the summer of 2000; the noise of the rotors made it impossible to hear the unauthorized speeches.

PRE-EVENT SEARCHES AND RAIDS (I, IV, AND XIV):

Preemptive strikes meant to reduce the size, force, and effect of a demonstration before it takes place. On the night preceding the Washington protest against the policies of the World Bank/IMF, city police raided an assembly point on Florida Avenue. Supposedly responding to a complaint about zoning and fire-code violations, the men in uniform confiscated the next day's handbills and parade puppets.

DENYING OF PERMITS (I AND XIV):

Although the denials frequently have to do with the applicant's political views or manner of dress, the stated reasons refer to the obstruction of traffic or the protection of the environment. When an activist organization sought permission to demonstrate on the Great Lawn in New York's Central Park last summer during the Republican National Convention, the municipal authorities refused the request on the grounds that a large crowd would harm the grass.

PAYING FOR PERMITS (I AND XIV):

Fees assessed by various municipalities for the privilege of staging a demonstration, sometimes accompanied by requirements that the protesters buy expensive liability insurance and defray the cost of any subsequent property damage.

PRE-EVENT PREEMPTIVE MASS FALSE ARRESTS AND DETENTION (I, IV, V, VI, AND XIV):

On April 15, 2000, in Washington, D.C., the evening prior to the World Bank/IMF protest, the police herded a crowd of people into a containment pen and then arrested more than 600 of them for no lawful reason. A fair number of the arrestees were confined in police buses for as long as eighteen hours without food, water, access to a telephone, a lawyer, or a toilet.

PREVENTIVE MASS FALSE ARRESTS AND DETENTION (I, IV, AND XIV):

Another form of arrest without probable cause, but imposed during the protest instead of as a preliminary precaution. In Washington on September 27, 2002, the police arrested 400 people in Pershing Park for "failure to obey" an unspecified order that was never given. Restrained in "flexi-cuffs" that tethered one wrist to the opposite ankle, the detainees could neither stand up nor lie prone.

SCREENING CHECKPOINTS (I, IV, AND XIV):

All bags become subject to search; large banners and signs must be given up to forestall their translation into weaponry. The more checkpoints that can be set up for no apparent reason, the more likely that the repeated demands for personal identification will promote the attitudes of boredom as well as fear, thus depressing the levels of enthusiasm and attendance.

FREE-SPEECH ZONES (I AND XIV):

Cages or fenced-off spaces in which citizens remain free to voice objections or display signs critical of a government official or policy. Anybody who expresses a contrary or insulting opinion beyond the designated perimeter is subject to arrest on charges of "disorderly conduct."

Whenever President Bush travels around the country to praise the freedoms for which America presumably is famous, the Secret Service sends advance scouts to set up the protest pens at a sufficient distance from the presidential speech or motorcade to preclude the possibility of the disturbance being seen or heard. At a Labor Day event in Pittsburgh in 2002, a retired steel worker was arrested for walking around outside the free-speech zone with a sign saying, "The Bush Family Must Surely Love the Poor, They Made So Many of Us." So also a man in Columbia, South Carolina, that same year found in the wrong place with a sign saying, "No War for Oil."

Prior to the Democratic National Convention last July in Boston, the National Lawyers Guild filed suit arguing that the free speech zone (a wire cage draped in black mesh) resembled "an internment camp." The presiding judge agreed that he couldn't conceive of a space that was "more of an affront to the idea of free expression," but, fearing for the safety of the convention delegates, he declined to shift or enlarge the venue.

SNATCH SQUADS (I, IV, AND XIV):

Also known as "extraction teams." Small bands of police, often wearing ski masks and full-body armor, that seize protesters chosen at random, hustle them into vans, and so disappear them from the next day's newspaper and television coverage. Among the other protesters present on the side-walk or the street the sight of their companions being beaten with clubs and abruptly carried off to points unknown serves as an inducement to proceed quietly and in good order to the next bus out of town.

LESS-LETHAL WEAPONS (I, IV, AND XIV):

Applications of excessive force that modern technology makes available to the science of persuasion. Among the most useful—tasers (stun guns capable of delivering 50,000 temporarily disabling volts of electricity), beanbag rounds (fired from twelve-gauge shotguns, effective at a range of sixty feet), concussion grenades.

INDICATIONS OF CRIMINAL ACTIVITY OR INTENT (I AND XIV):

Almost anything that comes to an arresting officer's mind. In October 2003, some days prior to the antiwar demonstrations scheduled for the following month in Washington and San Francisco, the FBI sent a memorandum to 15,000 local law-enforcement agencies citing as examples of potential criminal activity the use of tape recorders and video cameras as well as the wearing of sunglasses or headscarves presumably intended to protect the citizen against clouds of tear gas or pepper spray. In some jurisdictions cell phones qualify as "instruments of crime" and, if it so pleases the arresting officer, the possession of one constitutes a misdemeanor.

INTIMIDATIONS BY MEDIA (I):

By exaggerating the chance of violence at all demonstrations of any consequence or size, the authorities seek to reduce the number of people likely to attend. Thus the mayor of Philadelphia prior to the Republican Convention in July 2000—"we have got some idiots coming here . . . they are going to get a very ugly response . . ."; anticipating antiwar demonstrations in the fall of 2002, the mayor of Chicago informed the *Chicago Sun-Times* of the imminent arrival of "hordes of violent maniacs."

More often than not the authorities can count on the eagerness of the news media to promote the rumors of anarchists breaking into department stores. *New York* magazine last May dressed up its preliminary coverage of the Republican Convention with descriptions of the incoming protesters as a "Bush-hating nation of freaks, flash-mobbers . . . wannabe revolutionaries" certain to run amok around "vehicle checkpoints . . . manned with heavy weapons, dogs. . . ."

Two months closer to the dreadful days in question, the *New York Daily News* devoted its front page to the message that "Cops Fear Hard-Core Lunatics Plotting Convention Chaos," the prediction backed up on page 6 with an announcement from Police Commissioner Raymond Kelly to the effect that the lunatics "have increased their level of violence" and "are looking to take us on." By the time the delegates presented their credentials in late August the city had raised a regiment of 10,000 police officers, many of them armed with machine guns, to defend it against foreign bombs and domestic speeches.

If the demonstration proves to be either orderly or of a large enough dimension to lend force and point to its subversive sentiment, the news media take little notice of the event; the story shifts to an inside page, the appearance of the crowd reported as old, tired, irrelevant, and small.

Even in the best of times the dissenting voice doesn't attract a popular following, rarely walks on stage to the sound of warm and welcoming applause. In times of trouble the expression of contrary or unorthodox opinion comes to be confused with treason, and as a stop in the mouth of a possibly quarrelsome electorate, nothing works as well as the lollipop of a foreign war. The dodge is as old as chariots in Egypt, but ever since the September 11 attack on New York and Washington, the Bush Administration has had little else with which to demonstrate either the goodness of its heart or the worth of its existence. Let too many citizens begin to ask impertinent questions about the shambles of the federal budget or the ill-conceived occupation of Iraq, and the government sends another spokesperson to a microphone with another story about a missing nuclear bomb or a newly discovered nerve gas. Sometimes it's the director of the FBI, sometimes an unnamed source in the CIA, but always it's the same message—suspect your neighbor and watch the sky; say nothing that cannot be safely overheard; buy duct tape.

The instruction receives a high approval rating because it conforms to the character of a society badly crippled by the fear of nearly everything for which it can find a foreign name. Surrounding ourselves with surveillance cameras and security checkpoints, we learn to proceed on the assumption that everybody is guilty of something—if not plotting the murder of the American Olympic team then cheating the insurance or credit-card company, lying to both the doctor and the lawyer, carrying concealed weapons or emotions. The drug companies decorate our television screens with advertisements for miraculous cures of unspecified diseases; when stopping a suburban station wagon for a minor traffic violation, the arresting officers approach with their hands on their guns, as if expecting to find somewhere in the tangle of surfboards and tennis rackets a contraband Arab or a hidden kitchen knife.

When demanding a thumbprint or a urine sample, the agents of government ask the citizenry for its trust and respect, but who in his or her right mind can trust or respect nervous bullies who make arbitrary arrests and choose to look upon the American people as probable enemies deserving of suspicion and contempt? The authorities defeat what they say is their purpose; presenting a deal similar to the one offered the luckless peasants in Vietnam (i.e., to save the village by destroying it), they suggest that we preserve our liberties by placing them in administrative detention—temporarily, of course, for our own good, and in our own best interests. But the government doesn't lightly relinquish the spoils of power seized under the pretext of apocalypse. What the government grasps, the government seeks to keep and hold, choosing to forget that the health and well-being of the American democracy depends less on the swagger of its police forces than on the capacity of its individual citizens to muster the strength of their own thought. We can't know what we're about, or whether we're telling ourselves too many lies, unless we can see and hear one another think out loud. Democracy is by definition a work in progress, a never-ending argument predicated on James Madison's notion that whereas "in Europe charters of liberty have been granted by power," America has set the example of "charters of power granted by liberty."

If we mean to continue setting that example, we cannot expect the wisdoms currently in office to do the work on their own initiative. The report published by the National Lawyers Guild is very clear on the point. The country's law-enforcement agencies (federal, state, county, and city) routinely break the law, routinely mock, disregard, and trespass against both the letter and the spirit of the Constitution, but the Justice Department as administered by Attorney General John Ashcroft makes little or no attempt to restrain or discourage, much less prosecute, the crimes committed in the name of "the national security." Unless we find a jurisdiction in which to do so—and with it the courage to indict and punish rather than to regret and deplore—the country stands to lose the constitutional right to its own name.

October 2004

Straw Votes

To the extent that a citizen's right to vote is debased, he is that much less a citizen.
—Chief Justice Earl Warren

*I*f I lived in Cleveland or Detroit, my vote in the November presidential election might count for something in the eventual result; because I live in New York, it will count for nothing, as pointless as would be my vote for the next president of Uzbekistan or France. Roughly two thirds of the American electorate is similarly disenfranchised, and so it comes as no surprise that the autumn campaign season has brought with it a dense fog of slander in all categories of informed and uninformed opinion (Democrat and Republican, military and civilian, urban as well as rural and suburban) spewed forth by diminished citizens in both the red states and the blue who apparently take comfort in their feelings of resentment, alienation, and rage.

The American people don't choose the American president; the decision rests with the Electoral College, which, as was made plain four years ago in Florida, may or may not reflect the popular will.

The variance is deliberate, intended by the framers of the Constitution as a defense against the corruption of a federal legislature too easily bought and sold and as a check on the ignorant passions of an unlettered populace widely dispersed in what was still a wilderness. The Electoral College in the late eighteenth century recruited its members from among the most enlightened citizens in each of the states, men "free from any sinister bias," as well read as they were well traveled, admired for their "virtue," "discernment," and "information." By 1828 the theory of appointing wise counselors had given way to the practice of employing partisan stooges, but the Electoral College continues to exclude ordinary, run-of-the-mill Americans from the privilege of direct participation in the naming of the individual to whom they entrust the administration of their government. Which is why, together with everybody else in the country on Election Day, I won't vote for Senator John Kerry or President George W. Bush; I'll vote instead for thirty-one unknown persons pledged to a line on the ballot and chosen for no reason other than their reliably sinister bias.

The Constitution assigns to each state a specific number of electors, the size of the delegation based on population and representation in Congress—fifty-five for California, twenty-one for Illinois, three for Wyoming, etc. Acting as freight-forwarding agents for the plurality of votes cast in each state, the electors come together in fifty state capitols on the first Monday after the second Wednesday in December, and there, in the presence of flags and usually at noon, they transfer the entire allotment of their state's electoral vote to the candidate on the winning side of the percentage. They take no notice of, nor grant any standing to, the concerns, wishes, views, theories, or convictions on the losing side of the percentage. On November 7, 2000, nearly 3 million people in Florida voted for Vice President Al Gore, but because Governor George W. Bush received the plurality (by a disputed margin of 537 votes), the Electoral College awarded him every vote cast for

Gore and thus one more than the requisite majority of 270 in the Electoral College. Although in voting booths across the whole of the country Gore received 539,893 more popular votes than Bush, the single electoral vote, buttressed by the Supreme Court's decision to forbid a final recount in Florida, placed Bush in the White House.

In a new book published last summer under the title *Why the Electoral College Is Bad for America,* George C. Edwards III, a professor of political science at Texas A&M University, explains the anti-democratic procedure and result in December 2000 with reference to the Bible lesson given at Matthew 13:12: "For whosoever hath, to him shall be given, and he shall have more abundance: but whosoever hath not, from him shall be taken away even that he hath." The same words describe the method of the Bush Administration in both its foreign and domestic theaters of operation, and in our current state of political animosity and confusion, I've come across few books as timely or as relevant as the one in which Professor Edwards suggests that the country now finds itself confronted not only with an absence of a coherent national politics but also with a constitutional crisis.

Unable to see how a democracy can call itself a democracy unless everybody's vote is counted as equal, Professor Edwards sets out to prove that the Electoral College is both needlessly complex and inherently unjust. He informs his treatise with statistical tabulations (of election returns, presidential travel schedules, placement of campaign advertisements), with historical points of comparison (the elections of 1876, 1888, and 1960 discussed as foreshadowings of the election of 2000), and with firm refutations of the several contemporary pleadings put forward on behalf of the Electoral College as a necessary support of the two-party system. Other scholars in other rooms undoubtedly will quarrel with one or another of the professor's conclusions, but his principal lines of argument deserve extensive debate in both the news media and the Congress.

I. SOME VOTES ARE MORE EQUAL THAN OTHERS

Seeking to balance the interests of the larger states with those of the smaller states, delegates to the Constitutional Convention in 1787 devised the "Great Compromise" that apportioned seats in the House of Representatives according to the size of the state's population but assigned to each state, no matter how sparsely settled, two seats in the Senate. The deal was falsely named—not a compromise but a concession to the smaller states threatening to withhold ratification of the Constitution unless they received an equal share of America's newly acquired political inheritance. Writing in *Harper's Magazine* last May, Richard Rosenfeld described the consequences:

> In America today, U.S. senators from the twenty-six smallest states, representing a mere 18 percent of the nation's population, hold a majority in the United States Senate, and, therefore, under the Constitution, regardless of what the President, the House of Representatives, or even an overwhelming majority of the American people wants, nothing becomes law if those senators object.

For the same reasons that dictated the undemocratic organization of the Senate, the Electoral College under-represents large states and over-represents small states. As Professor Edwards points out, an electoral vote in *Wyoming presently corresponds to* 167,081 persons; an electoral vote in California represents 645,172 persons; which means that in a presidential election a popular vote in Cheyenne is the equivalent of four popular votes in Los Angeles or San Luis Obispo. The democratic faith in majority rule sustains and validates every other form of American election, but the election of the president takes place in an alternate universe.

2. THE IMAGINARY MAJORITY

Voters unaligned with the state's electoral vote play no part in the presidential election, and their voices disappear from the national political stage. The majority of the country's African Americans live in the southern states, their presence unremarked upon and their concerns unaddressed by either presidential candidate, because the southern states routinely deliver their electoral vote to the Republicans. Similarly, in the New England states, the electoral vote routinely goes to the Democrats, with the result that both presidential campaigns ignore the presence of conservative, socialist, libertarian, or independently minded voters in Rhode Island and Connecticut. Comparable to the demographic charts governing the sale and distribution of consumer products, the tally in the Electoral College presents a distorted picture of the American character and mind—too much emphasis assigned to the Christian fundamentalism in the South as well as to the secular humanism in the North, not enough *recognition of the diversity of opinion* in small towns as well as in large cities.

Election by a majority of states rather than by a majority of citizens excuses the candidates from the effort of talking to, much less attempting to convince or persuade, voters not already inclined to applaud most everything they say or promise. Four years ago neither George Bush nor Al Gore spent much time campaigning in Texas, California, or New York, three states that among them encompassed 26 percent of the American population but possessed no electoral votes deemed to be negotiable—the 32 in Texas on consignment to Bush, the 33 in New York and the 54 in California reserved to Gore. The two candidates in 2000 made a combined total of eighteen appearances in Wisconsin, thirty-one in Michigan, and twenty-six in Florida; neither of them appeared, not even once, in any of the eighteen states (among them Vermont, Oklahoma, Colorado, and Connecticut) regarded as dead letters and foregone conclusions.

The disproportionate investment of time and money in the so-called swing or battleground states is a measure of the degree to which the value of an American citizen has been discounted and debased. The presidential candidates don't address the American people simply as their fellow countrymen; they speak to an aggregate of interest groups and target audiences—Americans distinguished not by the fact of being American but by those of their ancillary characteristics that reduce them to a commodity: as a female American, a white American, a gay American, a black American, a Jewish American, a Native American, a swing-state American. The subordination of the noun to the adjective makes a mockery of the democratic premise, but it serves the marketing strategy of a campaign directed at the Electoral College, and it substitutes for a unified American or national interest an incoherent miscellany of state or special interests. Which is why the naming of the next president of the United States can turn not on a question important to the country as a whole but on a sentiment dear to the hearts of the Cuban Americans in the swing state of Florida.

3. CHICANERY

As of late September the twenty-one swing states that the presidential sales directors believed to be "in play" (among them Florida, Tennessee, Wisconsin, Minnesota, Ohio, Virginia, and Missouri) accounted for a total of 216 electoral votes (80 percent of the number necessary to win the White House). Four years ago in many of those states the difference between the winning and losing percentage of the popular plurality was extremely small (.22 percent in Wisconsin, .01 percent in Florida), which, if the political parties hold true to form in this November's election, should greatly improve the odds in favor of theft and fraud. Of the 115 million votes likely to be cast on Election Day, 36 million will be punched into direct recording electronic systems (DREs) that provide no paper receipt and no possibility of a recount independent of the computers under the control of local election officials and the corporations hired to vouch for the result. The state Division of Elections in Florida already has ruled illegal any attempt to recount votes subject to dispute.

By doing away with the Electoral College we wouldn't cure all the ills that currently afflict the American democracy—the state of political paralysis that follows from the two-party system, unrepresentative government in the Senate, etc.—but at the very least we might make a beginning. The reform would strengthen what James Madison once called "the vital principle of republican government"—one man, one vote, the will of the majority, the belief that all of us have an equal say in the matter. We prove ourselves citizens of a democracy not by our winning of elections but by our agreeing to lose elections. The deal is hard to make, and the consent of the governed not freely given, unless we think ourselves participant in the election. If my vote doesn't count, I have no stake in the outcome, no reason to accept responsibility for, or acquire knowledge of, either the good or evil done in my name by the government in Washington. A CBS/*New York Times* poll taken in May 2003 (i.e., twenty months after the collapse of the World Trade Center and eight weeks after the invasion of Iraq) discovered 38 percent of the respondents refusing to regard George W. Bush as the legitimate president of the United States, a finding that accounts for a good deal of the rancor in this autumn's election campaign.

Since the inception of the republic, a central theme in the American political story has been the one about the further broadening of the electorate and thus the further democratization of the Constitution. The intention has been abetted and approved by politicians as distant from one another in time and place as Presidents James Madison and Andrew Jackson, Senators Estes Kefauver, Hubert Humphrey, and Margaret Chase Smith. Five of the seventeen amendments added to the Bill of Rights since 1791 have expanded the electorate—the Fifteenth in 1870 (extending the vote to former slaves), the Nineteenth in 1920 (presenting the vote to women), the Twenty-third in 1961 (granting the vote to residents of Washington, D.C.), the Twenty-fourth in 1964 (prohibiting poll taxes), and the twenty-sixth in 1971 (welcoming voters to the polls at the age of eighteen). For the last fifty years the Gallup Poll has shown a clear majority of the American people in favor of a constitutional amendment dismantling the Electoral College; three years ago the poll reported "little question" on the part of the American public about going to "a direct popular vote for the presidency." In 1969 and again ten years later, the Congress nearly passed the necessary legislation, both attempts endorsed by strong majorities in the House but failing, narrowly, in the Senate.

Professor Edwards observes that for more than 200 years the country has survived the consequences of the variance between the popular and the electoral vote (the accession to the White House of Rutherford B. Hayes in 1876, of Benjamin Harrison in 1888), but I don't think that our luck is likely to hold for another three months, let alone another decade or century. In no prior election season can I remember talking to so many people who say, bitterly and seriously, that they intend to leave the country if their candidate fails to win the White House. Having lost faith in both the theory and practice of democratic self-government, they look upon the election in the manner of spectators at a bad play, amused or not amused by the whirl and spin of libel in the news media but believing themselves absent from the long and continuing American struggle (brave, dangerous, always against the odds) to secure a government of laws, not men.

Their indifference doesn't bode well for the country's future prospects, and maybe we should count ourselves fortunate if the

November election results in stalemate, suspicion, and dispute. The circumstance would oblige us to rediscover the purpose and meaning of democracy, to realign our political thought, and therefore the Constitution, with circumstances far different from those existing in the late eighteenth century, to elect as president a man or woman representing the whole of our national identity rather than the smiling face of a focus-group Caesar or Napoleon striking heroic military poses in a swing-state shopping mall.

November 2004

R & D

A country without its czar is like a village without an idiot.
—Russian proverb

*T*he documentary play *Guantánamo: Honor Bound to Defend Freedom* serves as a dispatch from the Cuban internment camps in which the American government currently holds captive several hundred presumed terrorists of Arab nationality and descent, and very early in the performance I saw last October in New York it occurred to me that I had been extended the privilege of watching a Pentagon experiment with laboratory animals. On the strength of the play's intelligence and from what I knew of its provenance (the script based on evidence gathered by British journalists and London civil-rights lawyers), I could assume that the association of ideas was deliberate and the irony intended. The principal actors appear as "detainees" dressed in orange prison uniforms and placed on a desolate stage furnished with a few tables and chairs, four narrow bunks, and four steel cages; for the most part silent and inert, they wait to be moved, like so many numbered mice, into another maze, tent, interrogation booth, or isolation cell. When permitted to speak about the circumstances of their arrest or the terms of their confinement, they use the words given in legal pleadings, press reports, and private letters from three British citizens, law-abiding but inopportunely Muslim, who found themselves among the herd of suspects rounded up by American military authorities searching the world for allies of Osama bin Laden during the months subsequent to the overthrow of the Taliban in Afghanistan.

The play doesn't quarrel with a civilized society's right to defend itself—if necessary, by whatever means come most expediently to hand—against enemies both real and imagined; neither does it doubt the possibility that at least some of the suspects brought to Guantánamo Bay provided information forestalling the destruction of a bridge in Maryland or the poisoning of a reservoir in Oklahoma. The play doesn't address the realpolitik of the war on terror; it considers the moral consequences—not the grand strategy of what President Bush defines as the "monumental struggle of good versus evil" but the brutalization of the participants on both sides of the interrogation, both ends of the rope. The actors speak as detainees who happen to be innocent—not surprisingly, because, as was made clear with the publication of *The 9/11 Commission Report,* our American intelligence agencies find it hard to distinguish one Muslim from another, to tell the difference between a jihadist mullah, an Iraqi politician, an Afghan warlord, and a Syrian bicycle thief. Among the inmates held at Guantánamo Bay for nearly three years, only four have been formally charged with a crime, apparently no more than twelve or maybe twenty guilty of some sort of a connection to Al Qaeda. The Supreme Court last June granted the detainees the right to inquire about the reasons for their imprisonment, but the questions of procedure have yet to be resolved. In what legal jurisdiction do the hearings take place, with or without advice of counsel, under whose rules of engagement? The shuffling and reshuffling of paper could continue for another three years; in the meantime the voices on the stage try to account for their presence in a void.

Their joint and corroborating testimony gathers its force from the gradual accumulation of small and wretched facts. No grandiloquent statements about man's inhumanity to man, no artful turns of phrase or plot, little else except plain narrative and the bearing of collective witness—one man arrested for no discernible cause while making a religious pilgrimage to Pakistan, another man shipped across the Atlantic Ocean in a freight container with a cargo of his dead and dying companions. Slowly and against its will, the audience comes to learn what sort of prison it is that America, "honor bound to defend freedom," has created in the image of its own fear on a distant Caribbean shore: a system of justice operating outside the bounds of national or international law, similar in its charter to one of the Enron Corporation's special-purpose entities, accountable to no authority other than the word of the American president and the whim of the American military command, which acts as warden, prosecutor, defense counsel, jury, judge, and, if deemed appropriate, executioner.

Classified as enemy combatants and therefore ineligible for the rights accorded prisoners of war, denied access to a lawyer or a writ of habeas corpus, the detainees fall into the category of a subhuman species available to experiment—kept in cages; exposed to deafening noise, unmuzzled dogs, extreme temperatures of heat and cold; stripped naked and searched for contraband in their teeth and anal cavities; deprived of food, medicine, water, and sleep; seldom allowed to stand or move unless shackled with the weight of chains.

First staged in London at about the same time that the photographs from Baghdad's Abu Ghraib prison were making the rounds of the print and broadcast media, the play attracted a good deal of notice in the periodical press, and before seeing the off-Broadway production

I was well enough acquainted with the Bush Administration's approach to suspected terrorists—in Afghanistan and Iraq as well as at Camps X-Ray, Romeo, and Delta, praised by Secretary of Defense Donald Rumsfeld as safe, healthy, and humane environments in "beautiful, sunny" Cuba—to know that the dramatization was faithful to the facts. What I didn't expect was the shift in perspective introduced by the notion of a Pentagon research project, and on leaving the theater I found myself wondering about the purpose of the experiments. What was it that our inspectors general were trying to find out, and why so many of them? Who was learning what from whom?

The large number of intelligence operatives (regular Army as well as CIA) sent to Cuba since the winter of 2002 to interrogate the same few hundred inmates suggests the need for a training facility where Christian officers and gentlemen might practice the art of extracting information from hardened infidels, improve their technique, overcome their feelings of revulsion and disgust. I can understand why it might be important to learn how to translate a scream in Arabic into a word in English, or useful to note the precise degrees of humiliation and degradation that a human being can be made to suffer without inducing insanity or attempted suicide, but how often must the experiments be repeated? Surely at some point the answers cease to be of interest. The research staff presumably looks for something other than fantastic guesses as to the whereabouts of Mullah Omar, and so I'm obliged to think that our apprentice Grand Inquisitors have more far-reaching questions in mind—not questions about the phobias of captive Muslims (how do they react to sexual insults, Big Mac cheeseburgers, and giant spiders?) but questions about the character and quality of free Americans.

If it is our intention to rule the world from the throne of military empire, how willing are the American people to tolerate or ignore, perhaps even to admire and applaud, the cruelties necessary to the maintenance of so great a glory? Is it possible to construct the moral equivalent of a toxic-waste dump in which to dispose of our sentimental squeamishness? If the government chooses to hang its prisoners by their testicles or thumbs, must the authorities in Washington anticipate objections from CBS News? From the Catholic and evangelical churches? From the Supreme Court? If so, how strong an objection, and can it be silenced with the antidote of fear? If a Marine colonel makes a mistake with an experiment involving two Syrian terrorists, a fishing boat, and a shark, will a feature editor at the *Washington Post* award the story seven paragraphs or one?

Different answers to the questions imply different versions of the American future, and as I considered the various possibilities in the light of the next day's newspaper reports arriving from Israel, Afghanistan, Russia, and Iraq, I noticed that it was hard to find much deviation between the reasons given by American generals for the bombing of Iraqi civilians in Fallujah and those given by Israeli generals for bombing Palestinian civilians in the Gaza Strip; or to make a clear distinction between Vladimir Putin's belief that too much freedom threatens the stability of the Russian state and the Bush Administration's aversion to any and all forms of constitutional law. Given the money and effort that the United States has assigned over the last half-century to the shaping of Russian and Israeli politics, it shouldn't come as a surprise that both countries now serve us as laboratories in which we might study various strains of anti-democratic government and cultivate socioeconomic organisms adaptable to the totalitarian climate of the new American imperium.

At present the most advanced research is being done in Iraq, much of it apparently directed toward a further and more complete understanding of the necessity for a state of perpetual war. George

Orwell identified the importance of the topic in the novel *1984*; Adolf Hitler conducted extensive field studies in both Eastern and Western Europe; America and the Soviet Union cooperated for forty years in a joint experiment with the waging of synthetic war, words substituted for deeds, prolonged artillery bombardment replaced with the constant threat of nuclear annihilation. The prior research need not be discounted or ignored, but in Baghdad at the moment we have access to a near limitless supply of laboratory specimens and a rare chance, literally God-given to President George W. Bush, to add to our store of knowledge.

The senior managers of the Bush Administration can be counted upon to acknowledge the truth of Orwell's dictum that "ignorance is strength," but will the Iraqi people verify the corollary finding that "freedom is slavery"? For how many weeks or months, and with what degree of religious zeal, will a true believer in the promise of Islam persist in his or her refusal to pledge allegiance to the American flag? To accept the word of Christ? Can our own soldiers be relied upon to decimate the ranks of our enemies with the same reptilian calm that the historians ascribe to the legions of imperial Rome? How quickly, and with what modifications to its assembly lines, can a nominally free press be converted to the production of weapons-grade propaganda?

The scale of the federal program in Iraq should yield results well worth the cost of the undertaking, but it cannot answer all the questions, and in some areas of related interest we will continue to depend on the experiments being performed in Israel, Afghanistan, and Russia. The Afghans test the hypothesis that an economy sustained by drug trafficking and a social order governed by a savage interpretation of the Koran can be presented to the world in the costume of democracy. Ever since its 1967 conquest of the Palestinian territories in Gaza and the West Bank, more desperately since the rising of the second intifada in September 2000, the Israeli government has been searching for new and improved techniques guaranteed to control the pestilence of a subject population. In the process it has developed lines of anesthetic reasoning, among them the theory of preemptive strike and precautionary assassination, to protect its own citizens against the pain induced by an overly active conscience. Many of the same arguments we have adopted as palliatives for our own states of anxiety, but we have yet to learn the secret of removing from the American body politic large numbers of people deemed undesirable, dangerous, or impious. The Israelis are fortunate to find every antisocial trait of character in the same enemy, and so their experiment with a wall neatly separating the just from the unjust might not prove immediately applicable to the American circumstance. Our society is too multifaceted, infiltrated by too many people of different races, colors, creeds, and sexual orientations. The work in Israel, however, deserves serious consideration and careful study. An alarming number of our most eminent political theorists and financial advisers foresee a soon arriving end not only to American democracy but also to the country's long-abiding economic prosperity. If their premonitions of heavy debt and chronic unemployment prove as well founded as their own off-shore bank accounts, how then do our ruling and possessing classes redistribute the presence of the no longer working poor?

Some of the more impregnable gated communities in the country's upscale suburbs already incorporate elements of medieval-fortress architecture, but they don't come fully equipped with floodlights, razor wire, and readily available armored vehicles; fences along the Mexican border from California to Texas are, in places, adequate to their purpose but not suitable to the terrain along the Canadian border in Minnesota and Montana. It's conceivable that we might wish to build model communities within the United States that combine the theory of the refugee camp at Khan Younis with the design of Camps X-Ray, Romeo, and Delta.

The lessons to be learned in the Russian laboratories have to do with the problems presented by a national economy fallen into the hands of thieves. During the long siege of the Cold War, Russia bankrupted itself in the attempt to compete with America's weapons industry and thus to earn promotion to the rank of superpower and the name of hegemon. The collapse of the Berlin Wall in November 1989 put an end to Communism, and within a matter of months a new class of arriviste oligarchs, schooled by American bankers in the science of high-end swindle, privatized what remained of the national wealth. Now comes the question as to whether they can keep the rewards of their entrepreneurial enterprise. The Putin government, increasingly authoritarian in character and method, seeks to repatriate the assets lost to the private sector. The fledgling system of representative government has been canceled by a return to czarism, the news media have been brought obediently to heel, and among the richest captains of Russian industry and finance quite a few have been forced to depart for London and the French Riviera. Mikhail Khodorkovsky, proprietor of the Yukos oil monopoly and a man believed to be worth $15 billion, currently occupies a jail cell in Moscow not much bigger than the ones reserved for the guests of the United States Navy in Cuba.

His situation is not without interest to our own fellowship of corporate burglars. How much is it possible to steal, and to what degrees of economic degradation and humiliation can the general population be exposed, before a virulent outbreak of a revolutionary virus obliges even the most venal and accommodating of governments to suppress the disease with the vaccine of despotism? Judging by the strong bipartisan support for the bill passed last October by both the Senate (69–17) and the House of Representatives (280–141) granting American business interests $137 billion in tax breaks, the day of reckoning is not yet come. Even so, prudence is a virtue, and it's always wise to know when the morning plane leaves for Zurich or Dubai.

December 2004

True Blue

*The spirit of resistance to government is so valuable on certain occasions that
I wish it to be always kept alive. It will often be exercised when wrong,
but better so than not to be exercised at all. I like a little rebellion now and then.
It is like a storm in the atmosphere.*
　　　　　　　—Thomas Jefferson

The London *Daily Mirror* published the result of last year's
American presidential election under the headline "How can
59,054,087 people be so dumb?" which was the same question that
on the morning of November 3 confounded every late- or early-
rising Democrat in Manhattan. If by noon I'd heard it asked in all
the tones of voice meant to express shock, disgust, bewilderment,
and shame, neither had I come across anybody equipped with an intelligible or ready answer. Among the company at lunch in a downtown restaurant catering to the literary trade what passed for
conversation consisted of little else except the exchange of stunned
silences. All present had been so certain that the election would go
the other way. How could it not? The American people might be
dumb, but were they also deaf and blind? Who but a lunatic or a
columnist for the *New York Post* could fail to see President George
W. Bush as a dishonest and self-glorifying braggart lost in the fog

of a quack religion. Surely the facts spoke for themselves. Under a pretext demonstrably false, the man had embarked the country on a disastrous and unnecessary war, mortgaged its economic future to foreign banks, assigned the care of the natural environment to the machinery certain to strip the land, poison the water, and pollute the air. What else did a voter need to know? Didn't people read the papers, look at the news broadcasts from Baghdad, wonder what had happened to their pension or their job?

So unforeseen was the calamity at the polls—the Republicans enlarging their majorities in both the Senate and the House as well as President George W. Bush winning a decisive plurality of the popular vote—that it was thought deserving of a higher order of politically scientific interpretation than ordinarily was to be found in a college civics class. Something heavy was afoot, something mystical or maybe criminal, and such is the speed of our modern system of communications and the fast-drying character of its instant wisdoms that within a single twenty-four-hour news cycle the tale of the Democratic Defeat was packaged in both an authorized and an unauthorized version. The mainstream print and broadcast media reported an election decided on "the moral issues"; the Internet blogosphere brought word of an election stolen by God-fearing thieves.

As was to be expected of the newspaper of record, the *New York Times* preferred the more hygienic text, and its op-ed page on November 4 offered no fewer than three commentaries on the separation of the country's spiritual and intellectual powers. The historian Garry Wills explained that the eighteenth-century Enlightenment had come to grief in the Pentecostal wilderness south of Chattanooga, that "many more Americans believe in the Virgin Birth than in Darwin's theory of evolution." Maureen Dowd deplored the politics of fear and intolerance with which President George W. Bush had recruited "a devoted flock of evangelicals" to the banners of holy crusade; Thomas Friedman discovered himself in a country undreamed of in the philosophy of Thomas Jefferson, where what has been lost is the distinction between church and state, where religion trumps science. The authors took no joy in their observations—possibly because their message was not much different from the propaganda delivered by Rush Limbaugh and the Reverend Jerry Falwell to the right-wing gospel crowd—but they testified to the existence of two Americas, one of them occupied by the virtuous souls in the great Midwestern heartland ("values voters," churchgoing and culturally conservative), the other inhabited by cynical apostates (nihilist at birth, often homosexual) trading foreign currencies and languages in the secular cities on the nation's seacoasts. Like it or not, the partition was encoded in the colors of the electoral map and therefore one to which we must pay heed.

Writing for the same op-ed page on November 6, Nicholas Kristof developed the evidence into a sermon. The time had come, he said, for the Democratic Party to find its way back to God. "I wish that winning were just a matter of presentation," he said, "but it's not. It involves compromising on principles." For the wayward politicians among his readers who might have lost their Bible in a Taiwanese bordello or the belly of a whale, he suggested a few first steps on the road to redemption—"Don't be afraid of religion"; argue theology with Republicans; "Hold your nose and work with President Bush as much as you can because it's lethal to be portrayed as obstructionist."

Similar instructions soon appeared on all the blackboards of the national news media—many fine words about the "healing" process binding up the wounds of partisan bitterness and strife—and on November 16 the chastened and recently reduced minority of Democrats still present in the Senate deferred to the sensibility of the nearest clergymen and named as their leader Harry Reid of Nevada. Non-obstructionist and much admired for his never having to hold his nose, Reid fit the description of a Democrat saved by Jesus—a teetotaling Mormon, a former Capitol Hill police officer, opposed to abortion, co-sponsor of the constitutional amendment deeming it a crime to burn the flag, careful to say nothing that anybody might remember, described by his colleagues as "strong as a new rope when he needs to be," as amiable as Mr. Rogers, the kind of guy who'll "make the trains run on time."

I don't doubt that the country is as rich in moral values as it is in apple trees, but I'm never sure that I know what the phrase means, or how it has come to be associated with the Republican Party, the Santa Fe Trail, or the war in Iraq. How is it moral for the President of the United States to ask a young American soldier to do him the service of dying in Fallujah in order that he might secure for himself a second term in the White House? Why is it moral to deny medical care to 40 million people who can't pay the loan-shark prices demanded by the insurance companies but to allow 12 million American families to go hungry in the winter? What is moral about an administration that never goes before a microphone to which it doesn't tell a lie?

Nor do I believe the sales pitch for the fruitful plains of Christian goodness lying to the west of the Mississippi River. When I read the advertisements distributed by the Heritage Foundation and the brewers of Colorado beer, I think of Tom DeLay (the House majority leader currently under threat of felony indictment, always quick to quote from scripture but quicker still to give or take a bribe), of Kenneth Lay (former chairman of the Enron Corporation, which relieved its investors of $60 billion and bilked the state of California of $2 billion), the young and entrepreneurial George W. Bush, who made his fortune by extorting it from the citizens of Arlington, Texas.

So also with the story about the conservative culture said to defend the faithful in small country towns against the wickedness of modern art and the sin of neon light, or the one about the sturdy forms of economic self-reliance that preserve the homespun country folk in Kansas, Nebraska, and Tennessee. Both stories are as false as the image of New York City as a sink of iniquity, or of the Republican Party as the friend of the common man. The red states live on the charity of the blue states, more abjectly dependent on government subsidy than a Harlem welfare mother or the owner of a California football team, and if network television and the supermarket press can be taken as a measure of the culture that sustains the heartland dream of heaven, it is a product of the pagan, not the Christian, imagination—Las Vegas the Garden of Eden, the miracle of the loaves and fishes outpointed by a winning number in the lottery, the 72,000 nymphs and fauns dancing to the tune of the pornographic websites nearer to hand than the 72 black-eyed virgins promised by Allah to the martyrs of Islamic jihad.

Like the romance of the American West, the virtue of the American heartland is a proposition floated by speculators on the literary and financial markets located in the nation's godless seaports. The nineteenth-century settlement of the trans-Mississippi west was promoted by New York railroad operators who promised the pioneers a land of milk and honey and sent their wagons forth into bankruptcy and a desert; the Hollywood entertainment industry performs a similar service for an audience wishing to imagine itself cast in a Frank Capra movie, guarded by comic book heroes, the happiest people on earth at play in the fields of the Lord, free to drink from the fountain of youth sold under the counter in an Ecstasy pill, over the counter as a prescription for Viagra.

I can understand why columnist Kristof might wish that President Bush's return to the White House was not "just a matter of presentation," but if I don't know how else to account for the result except as a matter of the cinematography and the sound effects, neither do I think it astonishing or deplorable. An American presidential election is a movie, usually a very bad movie, but the American public likes bad movies, and President Bush was more convincing in the role of Batman than was Senator John Kerry in the role of Flash Gordon. The production values are the moral values.

The movie playing in the mainstream media during the first week after the election conformed to the specifications of a major studio release along the lines of *Titanic* or *Lord of the Rings*; the one that opened in the blogosphere multiplex resembled a film noir independently produced by Quentin Tarantino. Rising from the depths of the cyberspatial void like flotsam from a sunken ship, the tumult of postings and emails brought forth long lists of numbers, ten-page attachments, rumors of Republican election officials as corrupt as Huey Long, and it was hard to know who or what was telling the truth, which websites could be trusted, and which ones were being operated by paranoid conspiracy theorists or Captain Nemo. Much of the testimony was anecdotal or so circumstantial as to be open to an inference precisely opposite to the one intended, but a good deal of it bore the stamp of reliable witness and incontrovertible fact. The fragments of a possible narrative could be found in what was known to have occurred in Florida—statistical anomalies, election laws configured to prevent any chance of a recount, the malfunction of easily abused voting machines, many voters denied access to the polls, large numbers of ballots spoiled or lost—and although the scraps of evidence didn't make the weight of an indictment on charges of either grand or petty larceny, they at least provided clues worthy of further investigation:

- A precinct in Franklin County, Ohio, possessed of only 638 voters awarded 4,258 votes to Bush.

- In forty-seven of the sixty-seven counties in Florida, Bush received more votes than there were registered Republicans.

- Of the 120,200,000 votes cast on Election Day roughly a third were processed by electronic voting machines supplied not by government but by private corporations, at least one of them (Diebold) controlled by a zealous partisan of the Republican Party who made no secret of his wish to bring victory home for the holidays. The software programs enjoyed the protection granted to commercial trade secrets.

- In three states that relied extensively on paper ballots (Illinois, Maine, Wisconsin) the exit polls corresponded to the final tally. In six states that relied extensively on electronic touchscreens (North Carolina, New Hampshire, New Mexico, Pennsylvania, Florida, Ohio) the discrepancy between the exit polls and the final tally invariably favored Bush.

- In ten of the eleven swing states the final result differed from the predicted result, and in each instance the shift added votes for Bush.

- Voters in six states, most particularly those in three Florida counties (Broward, Dade, and Palm Beach) reported touching the screen for Kerry and seeing their ballots marked for Bush.

- The electronic machines in Broward County began counting absentee ballots backward once they had recorded 32,000 votes; as more people voted, the official vote count went down.

- Exit polls in states equipped with verifiable paper receipts corresponded to the final tally; in states employing electronic touch screens the margin of difference between exit polls and the final tallies was as high as 5, 7, and 9 percent.

The credibility of the story adrift in cyberspace had less to do with the certainty of the numbers than with the character of the

Bush Administration. If we know nothing else about the government now returning to office in Washington, we know that it doesn't hesitate to cheat and steal and lie. Its family values are those of the Corleone and Soprano families, and thus in line not only with the heartland values of the nineteenth-century American frontier but also with the predatory modus operandi of our twenty-first-century business corporations; an administration capable of perpetrating the murderous fraud of Operation Iraqi Freedom almost certainly would count it as a loss of face if it couldn't further serve God's will by fixing a presidential election.

The other supposition lending credence to the tale told on the Internet followed from the indignant reaction to it on the part of the mainstream media. Under the front-page headline "Vote fraud theories spread by blogs are quickly buried," the *New York Times* on November 12 published a report that compensated for its lack of journalistic enterprise with a tone of mockery and disdain— "Weblog hysteria," "Experts soon able to debunk." The unnamed experts (professors at Harvard, Cornell, and Stanford) addressed only those suspicions that were most easily allayed; the more troubling questions they left unburied. Similar admonitions appeared in the *Washington Post* ("Ultimately, none of the most popular theories holds up to close scrutiny"), in the *Boston Globe* ("Much of the traffic is little more than Internet-fueled conspiracy theories"). The motions to dismiss were seconded by various spokesmen for the Kerry campaign, among them David Wade, quoted in the *Times* as saying that "I'd give my right arm for Internet rumors of a stolen election to be true, but blogging it doesn't make it so. We can change the future; we can't re-write the past."

We do nothing else except rewrite the past—in every morning's newspaper, every novel, poem, history book, interoffice memo, message posted on a refrigerator or the Internet. We inhabit the landscapes of our stories, and of the two best-selling fictions explaining the Democratic Defeat, I found myself more at home in the one about the robbery. Although not without its flaws, at least it was consistent with what I know of the country in which I was born and proudly count myself a citizen, the story vouched for in the writings of Henry Adams and Mark Twain, in line with the taking of the land from the Mexicans and the Indians, with the heroic scale of the government fraud embedded in the building of the transcontinental railroad, with the Teapot Dome swindle, the stockmarket collapse of 1929, the Internet bubble of fond and recent memory. An American story, true blue and fire-engine red. If the Democrats don't spoil it with a Bible and a flag, maybe they can regain the courage, traditional and culturally conservative, to steal the next election.

January 2005

Pilgrims of Hope

The Propaganda of The Faith is quite the largest, oldest, most magnificent, most unabashed, and most lucrative enterprise in sales-publicity in all Christendom. . . . By contract, the many secular adventures in salesmanship are no better than upstarts, raw recruits, late and slender capitalisations out of the ample fund of human credulity.
—Thorstein Veblen

*D*uring the weeks immediately subsequent to last year's presidential election the media crowd in New York promoted to the authority of holy writ the color-coded message on the nation's electoral map, and by the time the star of Bethlehem had been hoisted atop the Fifth Avenue Christmas tree it was next to impossible to find a newspaper sage or television talking head who doubted the wisdom of the three sublime power points.

I. America's precious store of moral value is for the most part located on church property in small towns west of the Allegheny Mountains and south of the Delaware River. Trace elements of Christian virtue still can be found in the seaboard settlements, but not in sufficient quantity to wash out the sins of envy and lust implicit in the success of the Hollywood entertainment industry, or the sins of pride and sloth embodied in the ruin of the Democratic Party.

II. The congregations of the faithful singing hymns in Gopher Prairie unfortunately lack the blessing of intelligence. Their stupidity doesn't detract from the perfection of their belief in Jesus, but it sets them up as easy marks for slick-tongued salesmen who come among them jingling with beads and trinkets and Republican campaign buttons.

III. It is the work of we happy and enlightened few here at the buffet table at the Metropolitan Club to negotiate a peace with honor between the country's spiritual and intellectual powers, to bind up the wounds of sectional bitterness and strife that separate the rival companies of the elect in the red states and the blue.

The story was as hard to swallow as the one about the Rapture, and well before Santa's elves completed the window decorations for the Disney company, I was beginning to make jokes I knew I'd be bound to regret. The notion of two Americas, one damned and the other saved, seems to me as nonsensical as most of the discussion of the country's "moral values"; nor do I choose to believe that everybody resident in Idaho and Nebraska is as dumb as Donald Rumsfeld. The supposition runs counter to my own observation over the last fifty-odd years as well as to my reading of the national character in the library of American history and biography and a fairly extensive acquaintance with the novels of Melville, Twain, Howells, James, Wharton, Dreiser, Faulkner, Cather, Anderson, Fitzgerald, Hemingway, O'Hara, and Roth. But to enter into arguments with media officialdom in Manhattan is a solemn undertaking. The manufacturers of the nation's seasonal truths are as self-regarding as the gentlemen of the White House bedchamber who applaud the comings and goings of the president's dog; they demand to see credentials bearing the stamp of executive authority, and it's never wise to arouse in them a suspicion of undue levity. Fortunately it so happened that in early December I had attended a symposium at New School University addressed to the writings of Thorstein Veblen, and in his remarks on "The Country Town" I found a purgative for the pompous additives being served with the Christmas cheer.

The discovery was as welcome as it was unexpected. In my own writing I've often cited Veblen's *Theory of the Leisure Class,* and I'd been asked to place in a postmodern context some of the bedrock principles of American business enterprise ("pecuniary decency," "the physiognomy of goods," "conspicuous consumption," "invidious comparison," etc.) that Veblen first set forth in 1899. The clarity of his thought doesn't date; we continue to live in a society that regards the possession of wealth as a meritorious act. Veblen points to a good many of the absurdities that follow from that superstition, and I was prepared to explain why and how the war in Iraq and the morass of our network television programming adheres to his "canon of honorific waste." But before being called upon to draw the parallel between the keeping of decorative parrots in nineteenth-century Newport and the building of useless weapons systems for the adornment of our twenty-first-century military deer parks, I had a chance to listen to Professor Sidney Plotkin, a political scientist on the faculty of Vassar College, who began his presentation with reference to Veblen's *Absentee Ownership and Business Enterprise in Recent Times*:

> The country town of the great American farming region is the perfect flower of self-help and cupidity standardized on the American plan. . . . The country town is one of the great American institutions; perhaps the greatest, in the sense that it has had and continues to have a greater part than any other in shaping public sentiment and giving character to American culture.

Over the course of the next twenty minutes Plotkin summarized the chapter in the book in which Veblen arranges "the perfect

flower" in the vase of his appreciative irony. The synopsis led me
to the unabridged text, which confirmed Plotkin's assessment of
Veblen as the earliest and most perceptive theorist of the cultural
and socio-economic presuppositions that light the paths of righteousness in what have come to be identified as the red states. Veblen attributes the location of any country town to a "collusion
between 'interested parties' with a view to speculation in real estate." The community's civic pride and municipal affairs thus con-
verge on "booming" and "boosting" the worth of the nearby land to
values as far "out of touch with the material facts" as the traffic can
be made to bear. In a word, and not to put too fine a point on it, a
communal scam, a plucking of pigeons and a shearing of sheep, the
humble antecedent of the Internet bubble, an early form of the Enron and WorldCom swindles, "an enterprise in 'futures,' designed
to get something for nothing from the unwary, of whom it is
believed by experienced persons that 'there is one born every
minute.'"

Because it was the business of the country town to turn a profit
on the sale of "divinely beneficial intangibles," the most adroit
publicity agents became its leading and most admired citizens.
What was wanted was not a man too heavily burdened with moral
or intellectual integrity but a salesman and a booster, a cockeyed
optimist loud with the promises of a fresh start and a second
chance, a good fellow along the lines of George F. Babbitt, Colonel
Beriah Sellers, or Ronald Reagan, the kind of man who inspired
confidence in tomorrow's rainfall or the price of next year's corn,
who knew how to get folks up and doing in return for a percentage
of the gate, rekindling the candles of their avarice with fanciful reports of the fortune in highway construction lying just below the
surface of the Centralia Swamp.

Veblen understood the country town as a "retail trading-station,"
getting what could be got out of the underlying "usufruct" of the
local farm population, and in the habits of mind fundamental to

the retail trade (*"suppressio veri, suggestio falsi"*) he located both the animating spirit of America's popular sentiment and the standards of conduct tailored to the true form of its moral code. The arts of American business were the arts of "effrontery, salesmanship, make-believe," all of them directed to the great and noble project of spinning gold from straw, and the man who would make a success of the enterprise did well to remember that "the beginning of wisdom in salesmanship is equivocation," which means that "when there is easy money in sight and no one is looking," he must be prepared to ignore the overly fine distinctions between run-of-the-mill "prevarication," "outright duplicity," and utter "absence of scruple." Self-preservation knows no moral law, and "solvency is always a meritorious work," not only because it "puts a man in the way of acquiring merit" but because it transforms him into a pillar of the community whose "opinions and preferences have weight" and therefore enable him "to do much good for his fellow citizens." The exact amount of good, of course, depends on the boomer's ability to strike a profitable balance between the outward and inward facets of his character, between the man in public (jovial, warm-hearted, relentlessly cheerful, forever innocent and pure of heart) and the man in private—cautious, opportunistic, cynical, predatory, at home and at ease in "the frame of mind of a toad who has reached years of discretion and has found his appointed place along some frequented run where many flies and spiders pass and repass on their way to complete that destiny to which it has pleased an all-seeing and merciful Providence to call them. . . ."

When Veblen published his reflections on absentee ownership in 1923, he was well aware that the country town had become "tributary" to the centers of credit in the corporate east, and he could foresee the changes of venue and method certain to accompany improved means of transport and communication, more efficient uses of advertising, increased resort to national brands and trademarks. But he had faith in the power of the American people to consume an ever-expanding abundance of goods, most of them as superfluous as they were overpriced. In a society that regarded an aptitude for acquisition as the chief measure of public approbation and private self-esteem, he knew that its most admired figures would come to be those who exhibited their prowess by inflicting injury, either by force or by fraud, both on their competitors and on the vast throng of their customers, "pilgrims of hope," with which an all-merciful Providence had replaced the guileless buffalo recently departed from the country's fruited plains.

Because Veblen accurately gauged the social and economic imperatives that endow the tasks of conspicuous consumption with the fervor of a patriotic duty, he knew that no matter what the shifts in circumstance, "the soul of the country town" would go triumphantly "marching on," ever "upward and onward" toward wider horizons and bulkier margins of net gain. As he foretold, so it has come to pass, not only in the American spheres of economic activity but also in the country's political arenas, where public men must pass the "test of fitness according to retail-trade standards"— i.e., replicate the qualities admired in real estate speculators who achieve a satisfactory division of labor between the outer child and the inner toad. The fitness test is bipartisan and all-American, passed with exemplary ease not only by George W. Bush in last year's election but also by nearly every other sitting member of the United States Congress as well as by Presidents Truman, Nixon, Johnson, Reagan, and Clinton.

To interpret a vote for President Bush as a sign of stupidity is therefore as wrongheaded as counting such a vote as a proof of devotion to the teachings of the New Testament or the writings of Edmund Burke. Whether set up as storekeepers on the banks of the Wabash or as symbolist poets in Brooklyn Heights, the American people appreciate, perhaps better than any other people on the face of the earth, the art of the con game, and they take for granted the slippages (carried on the books as a tax-deductible business expense) between the face and the mask. Fully conscious of the fact that the promises are false, the deals rigged, and the judge safely in the bag, they're conservative in the sense that they wish to protect and preserve the time-honored mores of the country town, and with them their own access to the ways and means by which it remains possible to screw the system. Reformist and left-leaning politicians they tend to see as officious inspectors intent upon closing the loopholes and removing from the grocer's scale the local thumbs heavy with the weight and fragrant with the soil of the sacred American heartland. So what if the vested interests in Washington reserve to themselves the larger portions of the apple pie? Such has been the practice of the vested interests since the heyday of Alexander Hamilton, entirely in keeping with the American spirit of things and not to be unduly frowned upon as long as the vested interests remember to leave enough crumbs on or under the table for the chambers of commerce in Sioux Falls and Medicine Bow.

A similarly conservative bias informs the country town's approach to religion, which Veblen describes both as a matter of "salesmanlike" cowardice and "expedient make-believe." Because the rural system of knowledge and belief can admit nothing that might "annoy the prejudices of any appreciable number of the respectable townsfolk," the would-be pillar of the community learns to sing along with any psalm or bouncing ball likely to purchase, at a reasonable cost, a large holding of community goodwill. Where is the percentage in the expression of a new idea, or the "harm done" (from a business point of view) "in assenting to, and so in time coming to believe in, any or all of the commonplaces of the day before yesterday."

Not only is there no harm done, but there is also the chance of a condo in paradise. Long odds, of course, may be no better than those available to the ticket holders in the New York or California lottery, but hey, "You gotta be in it to win it."

Understand the true American as a pilgrim of hope wherever he happens to be placed on the nation's electoral map, and it's no surprise that the dealers in the true religion package their "scheme of deliverance" in the manner of real estate speculations, or that the 5 steps to personal salvation lie along the same yellow brick road as the 12 steps to sobriety and the 237 steps to financial well-being. The dealers in "spiritual amenities" rely on the same natural resource of "credulous infatuation" as do the merchants of material comfort, and Veblen was especially admiring of "the publicity-agents of the Faith," who habitually promise much but deliver "substantially none" of the material advertised:

"All that has been delivered hitherto has—perhaps all for the better—been in the nature of further publicity, often with a use of more pointedly menacing language; but it has always been more language, with a moratorium on the liquidation of the promises to pay, and a penalty on any expressed doubt of the solvency of the concern."

The observation stands as a fair and fitting tribute to the blessed miracle of the Bush Administration.

February 2005

Democracyland

*A party which is not afraid of letting culture, business, and welfare go to ruin
completely can be omnipotent for a while.*
—Jakob Burckhardt

*A*s seen from New York through the screens of the print and
broadcast media during the weeks following last year's elections, the news from Washington seemed to promise the chance of
animated debate, maybe even strong and honest argument, when
the newly-minted 109th Congress assembled on Capitol Hill in
early January to take its collective oath of office. Such at least was
my supposition. Prominent Democrats were making it a point to
complain to equally prominent journalists about the muzzles placed
on the snouts of their integrity by the stage managers of Senator
John Kerry's failed presidential campaign—no loud objections to
the war in Iraq, no sarcasm spilled on the platitudes, nothing un-
civil about the patriot in the White House. The restrictions had
prevented them from saying what needed to be said about the deliberate and premeditated harm done by the Bush Administration to
the American people. But now that the senator from Massachusetts

had sailed merrily down the stream on his windsurfing board, the gag rule had been lifted, and they were free at last to speak the truth.

On the other side of the aisle the Republicans were even more upfront about their intention to tell a straight and candid story. Emboldened by their November victories in both the presidential and congressional elections, the party's fuglemen were touting their plan to destroy, in all its liberal tenses and declensions, the hated remnants of Franklin Roosevelt's New Deal. At his triumphant press conference on November 4, President Bush made no attempt to conceal the ferocity of the forthcoming preemptive strike. "I earned capital in the campaign—political capital," he said, "and now I intend to spend it." By Christmas what had become known as the "Bush Agenda" encompassed the privatization of the Social Security system, a reformulation of the tax code in such a way as to provide more money for war and law enforcement, less money for the undeserving poor, the nomination to the Federal Appeals Courts of judges apt to find legal precedents in the books of the Bible rather than in the Articles of the Constitution, more laws limiting the freedom of individuals, fewer laws restraining the freedoms of property.

The lines thus so clearly drawn on the sand tables of the media raised the hope that somewhere in the Capitol or its vicinity a traveler from the provinces was surely bound to come across restless stirrings of parliamentary debate, bold expressions of impolitic dissent, some reason to believe that the forms of democratic government we see pictured on the postcards exist in substance as well as name.

The expectation was short-lived. Under a pale winter sun on the morning of January 4, the Capitol grounds resembled a military encampment. No leaves on the trees, few birds in the sky; the spacious vistas interdicted in all directions by armed men in black uniforms—police at the perimeter barricades, police on motorcycles, police drifting overhead in helicopters. Standing for an hour in the long line of citizens waiting to submit to the security procedures, I understood that it was neither the time nor the place to recite the Gettysburg Address. The impression was that of a medieval walled town preoccupied with its own weakness and fear, and well before I reached the last fortified checkpoint I knew that the notion of a government by the people, for the people, and of the people wasn't the kind of thing likely to meet with the approval of the metal detectors.

The bulwark of suspicion was reinforced by the heavy police presence inside the Capitol, every thirty yards another man in uniform asking for an identity, the official attitude similar to that of customs inspectors inclined to look upon lost luggage and Arab tourists as disguised weapons of mass destruction. It was another hour before I'd found my way to the office in which I filled out a form, sat for a photograph, and so received the press badge that allowed safe passage through the checkpoints. Still obliged to empty my pockets, of course, but no need to remove my shoes.

At noon in the Senate Chamber almost the whole of what its one hundred members like to call the "greatest deliberative body in the world" gathered for the swearing-in of their newly elected companions (nine for a first term, twenty-five for additional terms), and as they stepped forward four and five at a time to swear the oath of office in the presence of Vice President Dick Cheney, I was struck by the ways in which they looked so much like one another. The media flood the nation's editorial markets with testimonies to the piebald character of the American democracy jumbled together from a wonderful diversity of colors, creeds, and cultural dispensations, which is a swell story, but in the United States Senate not one visible to the naked eye. The press gallery affords a close and well-lighted view of the Chamber, and with it an occasion to study the collection of faces as if they already had become portrait busts in

Statuary Hall. Even at the privileged distance of less than twenty feet it was hard to imagine any of the members present—middle-aged and comfortably settled in their flesh, white, wearing expensive suits, glad to be here in Tampa for the golf outing—finding the time to write his or her own speech, much less taking the trouble to read through the 2,858 pages of the Federal Budget that distributes an annual appropriation of $2 trillion. Nothing in their manner suggested a shred of difference in their preconceptions and modus operandi. Red state, blue state; Old Testament, New Testament; popular assembly, oligarchical junta—why argue the details as long as everybody knows how and when to count the money?

The swearing-in ceremony was accomplished in less than an hour, Senator Kerry notable for his absence. Senator Bill Frist of Tennessee, the Republican majority leader, then delivered a speech welcoming "everyone here and everyone watching at home . . . to this historic first day of the 109th Congress." Although never a man known for his oratory, Frist did his best to impart to the words the flourish of high flown sentiment accompanied by stately gestures in the manner of Henry Clay—"My colleagues . . . [we] are the stewards of this ancient and yet still living and thriving tradition . . . The American people—and indeed the people of the world—look upon this Capitol and those of us who serve here for inspiration and leadership and unwavering devotion to our common cause. . . . My fellow Senators, you are all honorable men and women. . . . God bless you. . . ."

The effect was disconcerting because by the time Frist arrived at his second paragraph, hardly anybody remained in the chamber (two stenographers, the clerk, and Senator Harry Reid of Nevada, the minority leader obliged to speak next), which meant that Frist was addressing what I'm afraid he mistook for his eloquence to nobody else except the cameraman recording the event for C-SPAN and posterity.

Reid's speech consisted of a tribute to Frist ("one of the most

prominent transplant surgeons in the country"), a senseless and dis-
jointed anecdote about his father rescued from certain death in an
Arizona mine shaft, and a vow to seize "the bipartisan opportunities" available to this new Congress looking "to the future with a
greater day, a nicer day, a more pleasant day ahead." Unlike Frist,
Reid made no attempt at rhetorical grandeur, content to read his
prepared text straight into the camera lens in the flat voice of a real
estate agent bored by his own sales presentation.

Much of the legislative business brought before the Senate and the
House of Representatives later in the afternoon continued in the
same spirit and tone, for the most part consisting of routine measures appointing committees, establishing rules and procedures,
expressing sympathy for the victims of the Christmas tsunami in
the Indian Ocean, mourning the loss of American soldiers in Iraq.
A similarly dull calendar on Wednesday and Thursday gave me the
chance to seek out a number of Democrats whom I admired for
what I'd seen of their reflections in the news media, among them
Senator Byron Dorgan of North Dakota, Congressmen Edward
Markey of Massachusetts and Henry Waxman of California, Con-
gresswoman Nancy Pelosi, also of California, the minority leader in
the House.

As forthright in my bias as the talking-heads at Fox News, I began each conversation by expressing the hope that somehow the
Democrats might find the ways and means with which to counter
the Republican motion to reconstitute the United States in the image of Mexico. None of the four respondents quarreled with the observation that what was now at risk in the 109th Congress was
nothing more nor less than the principle of democratic government, which, given their constituencies and voting records, wasn't
surprising; what was surprising was both their sense of ineffectual-
ness and their agreement as to the obstacles standing in the way of

the animated debate that I'd been pleased to think possible when talking to myself in New York.

"It's truly amazing," Waxman said, "that so many people still think that this place is on the level." He explained that ever since the Republicans gained the majority in the House in 1994, the House leadership had been changing rules—eliminating the possibility of debate when one of their own bills comes to the floor for a vote, routinely giving the Democrats as little as twelve hours to read 800 pages of small and stupefying print. No Democrats were invited to the House and Senate conference considering last year's intelligence bill; nor were any Democrats allowed to propose an amendment to the medical prescription bill. Congressional requests for information from the executive agencies of government— from the Pentagon about the cost of weapons, from the Justice Department with regard to its policies on torture and the detention of "enemy combatants"—may or may not receive the courtesy of a reply. In the absence of answers to their questions, Congressional Democrats lately have been forced to file lawsuits in order to discover how the government for which they're held responsible conducts itself behind soundproofed doors.

As an instance of the strong-arm methods deployed by the Republican leadership in the House, also of the majority's contempt for the due process of law, Nancy Pelosi mentioned the new rule, passed with no chance of amendment on the first day of the 109th Congress, that rendered meaningless the name and purpose of the House Committee on Standards and Official Conduct. Evenly divided between Republicans and Democrats, the Committee henceforth will investigate no charge of moral or financial wrongdoing unless at least one of the Republicans present provides the enabling vote, an event as unlikely as a descent on Washington by the armies of Napoleon.

"These people are shameless," Pelosi said, "arrogant, petty, shortsighted." Representative Markey chose stronger words to express the same meaning. "They do as they please," he said. "They wish to wipe us out."

More than once while listening to the several confessions of parliamentary weakness, it occurred to me that our elected representatives of government construe themselves as having been reduced to the peonage of journalists. Dorgan had reformulated the Democratic Policy Committee to hold hearings meant to advertise the malfeasance of the Bush Administration—hearings about the subversion of the Social Security System, about the Halliburton Company's failure to account for the $10 billion that it had either stolen or buried in the deserts of Mesopotamia—but because the committee lacked subpoena power as well as legislative footing, it would depend for its effect on the whim of the news media. Would CBS News send a camera, or the *New York Times* a reporter? Waxman likewise presented himself as a mere gadfly, doomed to convene press conferences in the hope that somebody would accept the invitation. Markey described Congress as a "stimulus-response institution," taking its cues from the expression of public outrage that maybe could be incited by the circulation of e-mail and messages posted on the Internet. "We must capture the words," he said, "convert issues into melodrama—children dying, mothers weeping. The coin of the realm." Not wishing to discount Markey's last, best hope for Senator Reid's "nicer and more pleasant day ahead," I refrained from saying that the coin was counterfeit, that to think the media blessed with courage, conscience, or convictions was to build one's house on mud and sand.

The point didn't need belaboring because it was made clear the next day when the Senate Judiciary Committee briefly examined the qualifications of Alberto R. Gonzales, the White House counsel, to serve as attorney general of the United States. The nominee showed himself to be a man of little principle and less integrity, a clever eunuch in a corporate harem, grinning and self-satisfied, unwilling to give a straight answer to questions about the part he played in the drawing up of the memoranda for President Bush that referred to the Geneva Convention as "quaint" and "obsolete," and defined torture as "only physical pain of intensity akin to that which accompanies serious physical injuries such as death or organ failure." When asked for specific recollection of documents that the White House refused to release to the committee, he dodged behind the phrases "I don't recall . . . I don't remember . . ." Obviously, Senator, "his [President Bush's] priorities will become my priorities. . . ."

Nor did the Democratic members of the committee hold the judge accountable either to the facts or to the tests of scorn and ridicule. Senator Edward Kennedy and Patrick Leahy expressed concern, even tried to make sense of the bowdlerized record, but neither of them were willing to risk their depleted store of political capital on a bet already lost.

Among the 250-odd people crowded into the hearing room in the Hart Office Building, the majority were reporters come to see and not to tell. Failing to find the stuff of melodrama (children dying, mothers weeping), they turned the story into a corporate press release to which their editors affixed headlines signifying nothing—"GONZALES DEFENDS HIS WHITE HOUSE RECORD" (the *Washington Post*), "GONZALES SPEAKS AGAINST TORTURE DURING HEARING" (the *New York Times*).

Or, in plainer language, power is as power does, and if it's accountable to no law other than its own, well then, dear reader, at least you've seen the pictures and heard a government spokesman say that America never tells a lie. What else do you expect? Maybe a piece of marble quarried from one of the Capitol's portrait busts, or possibly a small square of glazed tile cut from the flooring of the Rotunda. A souvenir. Something to remind a tourist of what was once a great republic before it lost the war on terror.

I couldn't have guessed at the scale of the defeat until I came to Washington with the hope of proving it a dismal rumor. But except as proofs of fear and weakness, how else to interpret the practice of torture as state policy, the nervous habit of official secrecy, the military entrenchments around the Supreme Court and the Capitol?

Shortly after noon on Thursday the police locked down the building for twenty minutes—nobody allowed to enter, nobody permitted to leave—while one or another of the government's praetorian guard units (Secret Service, possibly the Wyoming National Guard or elements of the 82nd Airborne Division) cleared the grounds and the nearby streets for the arrival of Vice President Cheney and his entourage of motorcycles. On the west front of the building workmen were setting up the defenses (tactical, strategic; ancient, modern, and medieval) designed to protect Bush's inauguration later in the month from so vast a host of enemies (real and imagined, foreign and domestic) that the list of suspects amassed by the Department of Homeland Security was said to run to almost as many pages as were to be found in the Pentagon's library of plans for the domination of five continents and seven oceans. The news media already were chattering about the magnificent display of vigilance scheduled for January 20—8,500 uniformed officers securing the perimeter of the parade route on Pennsylvania Avenue, thirty-one checkpoints, dogs trained to sniff out explosives, sniper teams on rooftops, patrol boats in the Potomac River, monitors sensitive to poisonous substances in the atmosphere, political protesters confined to cages well out of the sight of Peter Jennings.

Walking down and away from Capitol Hill on Thursday afternoon I didn't notice any riflemen practicing their aim on squirrels or pigeons, but I could see the construction gangs extending the fortification of the West Steps, and in the distance beyond their cranes and pulleys, the Washington Monument and the Lincoln Memorial. Maybe it was a trick of the fading light, but instead of calling to mind the strength of the American spirit, the two landmarks at first glance reminded me of stage props for a television news show or a Hollywood movie, conceivably for a theme park Democracyland where, twice on weekdays and three times on Sunday, top quality high-school marching bands perform that well-known and much beloved musical number, "Land of the Free and Home of the Brave." The thought was not one that I could slide through the checkpoints at an Inauguration Ball, and I figured that the sooner I got back to New York the better my chance of finding an American political idea not so frightened of a future in which most of the days were apt to be neither nice, nonpartisan, nor pleasant.

March 2005

Stitches in Time

War is God's way of teaching Americans geography.
—Ambrose Bierce

*T*he train from Paris to Brussels passes through fields sown for 2,000 years with the seed of war, and on the way north last February 1 to the opening sessions of this year's European Parliament, I was reminded of the brightly beribboned armies—Saxon, Roman, Norman, English, French, Spanish, Austrian, German, and American—that had enriched the soil with the compost of military glory. Because I tend to read history with no particular beast or century in mind, I lose sight of the chronology in the mêlée of medieval chivalry and the movements of Renaissance cannon, which is why, looking out the window of the train at the flat and barren landscape presumably planted with winter vegetables, I didn't know whether to assign the summer harvest to the decay of a Nazi *Sturmbannführer* or to the disappearance of a Plantagenet king.

To the best of my knowledge, the European Union doesn't provide an agricultural subsidy for carrots emerging from the body of

Charlemagne or the blood of Wilfred Owen, but had the topic been proposed for discussion later that afternoon in the appropriate committee, I would have expected the Mayor of Amboise to submit a report noting the contribution made by the St. Bartholomew's Day Massacre to the quality of the dairy products in the valley of the Loire. It was the kind of question that suited my purpose in Brussels, which was to compare the European practice of democratic government with what I had seen a month earlier on Capitol Hill in Washington; if I could place the first week of February in the European Parliament against the backdrop of the first week in January at the American Congress, maybe I could come up with a best guess as to which of the two variations on a theme by Thomas Paine offered the better chance of an answer to the dilemma of a new millennium.

I didn't have far to look for the first of the differences between Washington and Brussels. The European Parliament occupies a suite of post-modern buildings in the upper part of the city, the arrangement of glass, steel, and polished stone along the lines of a California resort hotel, but at none of the several entrances were the police procedures as nervous as those in place on Capitol Hill—no dogs, no men in uniform armed with assault rifles, so little emphasis on security that I was permitted to smoke a cigarette in the third floor café favored by the Czechs and the Finns. The European politicians apparently weren't as frightened of Arabs or tobacco as were their American counterparts. Death was an old story, and only once in three days did I hear anybody mention the word "terrorism." Enrique Barón Crespo, a Spaniard and a socialist serving as chairman of the Parliament's Committee on International Trade, referred to the subject not as a matter of concern but as a commentary on the American fear of the future. A member of the Spanish government in the 1980s, Crespo on more than one occasion had been marked for assassination by the anarchist cadres then active in Barcelona and Madrid, and he had learned that it's no good living in a state of constant hysteria. "What's the point?" he said. "You could as easily be run over by a bus."

Most of the American foreign-policy experts who write about the EU dwell on its bureaucratic meddling with the Promethean engine of the free market, its laughable attempts to conduct itself in a manner befitting a global superpower. How to rid the world of tyrants or carry the gift of freedom to all four corners of the grateful earth when the so-called statesmen in your midst squander their days worrying about the size of a Portuguese shrimp?

The sarcasms never lack for supporting anecdote—tomatoes held to a uniform standard of color and roundness, every cow in France awarded unemployment benefits and a pension—but they fail to account both for the current measures of European prosperity and for the degree to which the European Parliament over the last thirty years has increased the weight and extended the reach of its legislative authority.

The received wisdoms were put to rout on the afternoon of February 2 during the newly elected body's first debate on the questions it was likely to confront over the course of the forthcoming year. Having been briefed on the mechanisms that assure the legislative and therefore democratic control of both the European Commission and the European Union, I could understand enough of what was being said to know that the Parliament had evolved into a far more formidable assembly than the one that I had last seen in Strasbourg in 1981. In what was then a theater of the absurd, a majority of the members had been elected from constituencies so small and so distant from the operative agencies of political power that their speeches were indistinguishable from slogans fit for waving in a protest march—defiant, implausible, romantic, loud. Other than myself—a tourist who happened to be passing through Strasbourg on the road to Mainz—the only people present were the four speakers waiting a turn to astonish with their sound and fury the silence of an empty room.

Thirty years later in Brussels the scene was as lively as the foreground in a painting by Brueghel the Younger—all the leaders of the Parliament's several voting blocs (Conservative, Socialist, Green, Liberal Democratic, Communist) gathered in a large amphitheater descending toward a dais staffed with officers of the European Commission (the EU's version of an executive branch of government), a jostling of television cameras, the importance of the proceedings attested to by both the front-page news story and the lead editorial in that morning's edition of London's *Financial Times*. Here then in a fanfare of press releases was the "Lisbon Strategy," the revisionist program of economic growth and development intended to confer on Europe a surplus of profit margins similar to those posted by Wal-Mart and Warren Buffett. As presented by José Manuel Barroso, the Portuguese president of the European Commission, the Lisbon Strategy envisioned a miracle of "increased productivity," "privatized" public services, "competitive" labor markets. First the money, Barroso said, then the comforts.

The order of priority didn't comfort the majority of the politicians in the room, chief among them Daniel Cohn-Bendit, president of the Parliament's Greens, who told Barroso that he was embarking on a "fool's errand," chasing an American dream of development that was as heartless as it was stupid. For the better part of three hours, the discussion was sufficiently wide-ranging to allow for the chance of comparison with the political arguments that take place on Capitol Hill, most obviously in the number of questions asked that seldom reach the floor of the United States Senate—global warming accepted as scientific fact rather than as tendentious theory, concerns, even on the part of politicians associated with the parties of the right, as to the wisdom of easing the environmental restrictions on the European chemical and automobile industries in the hope of a higher number expressing the rate of economic growth. Although I may have misheard the translation from the Swedish, on the latter point I think somebody said that probably it wasn't a good idea to fatten a banker's goose in return for a mess of poisoned fish.

As the Parliament over the years has come to represent more European countries (nine in 1979, twenty-five in 2004) so also it has enlarged its capacity to make politics—the power of the purse over the EU's annual budget of 100 billion euro, the power to co-decide legislation introduced to the national assemblies in the member states, the power to set rules governing the common lines of transport and communication, the power to write a European constitution and with it the chance to make something new under the sun—a civil society formed by a community of opinion not rooted in the egoisms of nationalist sovereignty and populist sentiment.

The diverse character of the 732 members (engineers, journalists, factory owners, scholars, musicians, labor unionists as well as the customary lawyers and professional politicians) provides a high enough quotient of disagreement to encourage genuinely democratic compromise. Elected in their own countries to represent an intellectual rather than a regional or economic interest, the members (30 percent of them women) can afford the risk of expressing a thought that hasn't been stuffed into their heads by a pollster or a K Street lobbyist.

Together with the American journalists who make fun of the EU, a good many people in Europe assume that the regulation of their commerce is in the hands of impotent bureaucrats dependent for their knowledge of farm animals on the designs of Pablo Picasso. The belief is popular but false. The attempts at coherence in what are known as "matters of European competence" derive from a legislative rather than an administrative authority, and over the course of the afternoon I had the time to reflect on the differences between the indices of a European and an American success—$20 million apartments on Fifth Avenue and the world's most wonderful air force as against the freshness of the bread on a Roman table, the absence of machine guns at the border crossings between Austria and Switzerland, free admission to a hospital, the certainty of an intelligent education. For everybody who can afford the price of a Harvard diploma and a pet politician, America is a very nice place to live; for people not so fortunately situated, America is fast becoming a brand name pasted on a bad movie or an empty box.

Various sets of statistics establish the exchange rate between the currencies of the private and the public good—one American adult in every five living in a state of poverty, as opposed to one in every fifteen in Italy; the quality of America's health-care services ranked thirty-seventh among the world's industrial nations; productivity per hour of work lower in the United States than in Belgium, Norway, the Netherlands, Germany, and France; Europe in 2003 giving $36.5 billion to other countries in need of development money, while a third of that sum was forthcoming from the United States; the disparity in the incomes of a CEO and a common laborer standing at a ratio of 475 to 1 in America, 15 to 1 in France, 13 to 1 in Sweden. The less abstract comparisons between the standards of living show up on postcards (the look of the architecture, the taste of the food and drink, etc.), but I think it worth noting that in the arena of foreign trade the American export of advanced-technology products declined by 21 percent in 2004, as opposed to its rising export (up by 135 percent) of scrap and waste. The numbers serve as a gloss on our current accounts deficit ($164 billion) and the fall in value of the dollar over the last few years (nearly 30 percent) when fixed against the euro.

The American notions of success, spiritual as well as material, congregate around the altar of the divine self. The display or consumption of the best of everything available online and in stock (hair products, beachfront property, religious guidance, domestic servants, and foreign oil reserves) serves as proof of salvation in both this world and the next. Given the ever larger number of people north and south of the equator seeking a share of the resources no longer as plentiful as the buffalo on the old Oklahoma frontier, the American song of perfect happiness begins to sound off-key in the political theater of the twenty-first century.

Cohn-Bendit made a corollary point with reference to the Bush Administration's messianic foreign policy. I'd remembered him as an ardent voice of protest during the student rebellion in Paris in 1968, and having been impressed by his remarks on the Lisbon Strategy, I stopped by his office two days after the parliamentary debate to ask a number of questions about the politics of the EU. He chose instead to talk about President Bush telling an audience of military officers in Washington on the night before his second inauguration that as Americans, he and they had "a calling from beyond the stars to stand for freedom."

"Where is that?" Cohn-Bendit said. "Beyond the stars. What do you say to people who think that freedom is given by God, not made by men? You can say, 'Are you mad?' but then, please tell me, what do you say next?"

The question touched on what over the course of three days I came to regard as the chief difference between the usages of the democratic idea in Washington and Brussels. On Capitol Hill as at the White House and the Pentagon the idea comes wrapped in the swaddling cloth of holy writ, accepted on faith and formulated as doctrine—Thomas Jefferson born in a manger, the Constitution brought from afar, by James Madison and Alexander Hamilton riding to Philadelphia on camels. The members of the European Parliament know something of history, and whether I was talking to the Baroness Emma Nicholson, a British peer whose ancestors had come to England with William the Conqueror, or to Bronislaw

Geremek (member from Warsaw, jailed as a political dissident during the early days of the Polish uprising in 1981, subsequently the Polish foreign minister), the conversations invariably took place in the presence of both the near and distant past. Europe in the twentieth century had twice attempted suicide, and nobody was eager to repeat the performance. The Nazi and Napoleonic dreams of empire were still as vividly in the news as President Bush's second inaugural announcement that America had lit "a fire in the minds of men. It warms those who feel its power, it burns those who fight its progress, and one day this untamed fire of freedom will reach the darkest corners of our world."

The difference between the sacred and the secular defenses against the arrows of outrageous fortune is the difference between certainty and doubt, and if given a choice of allies I'd rather take my chances with people who know that poverty, disease, and the degradation of the environment are greater weapons of mass destruction than the bombs coveted by Donald Rumsfeld and Saddam Hussein, that what saves us in the end is the force of the human imagination, not the armies sent by a deified emperor in Vienna or Rome. "The war on terror," Crespo had said, "is in the head; you don't win it with tanks."

I had come to Brussels with the hope of encountering a less terrified response to the storm of the world than the one I'd encountered in Washington, and despite the many failings of the European Parliament—the going to Strasbourg once a month to cast its ballots, the labyrinth of bureaucratic inertia, etc.—I took heart from its willingness to learn from experience and to employ the tools of constitutional government that in America have become museum pieces, to find its security in the health, courage, and intelligence of its citizens rather than in the four-color photographs of invincible aircraft carriers, to understand the democratic idea not as a projection of power but as an expression of liberty.

"We have made Europe," Geremek had said, "but how do we

make Europeans?" What he had in mind was a civilization in place of a fortress, and although his question was unanswerable, it seemed to me somehow better matched to the complexities of the twenty-first century than the ones that get asked in Washington about the size and throw weight of the President's codpiece.

April 2005

The Wrath of the Lamb

> *The theologian may indulge the pleasing task of describing Religion as she*
> *descended from Heaven, arrayed in her native purity. A more melancholy duty is*
> *imposed on the historian. He must discover the inevitable mixture of error and*
> *corruption, which she contracted in a long residence upon earth, among a weak*
> *and degenerate race of beings.*
> —Edward Gibbon

*A*t a press briefing in Washington early last March, the National Association of Evangelicals declared its intent to lend a
hand in the making of an American politics faithful to the will and
"abundant wisdom" of God. Taking into account the many and
atrocious proofs of God's incompetence as a politician, the announcement in less troubled times might have been seen as a
clownish hallucination or a bleak postmodern joke, but the association numbers its membership at 30 million exalted souls, one
fourth of the nation's eligible voters, and so the news media in attendance were careful not to laugh when the telegenic pastors,
smooth-faced and smiling, distributed a twelve-page manifesto for
a Bible-based public policy entitled "An Evangelical Call to Civic
Responsibility." The words were pretty enough, but to read the
document with any care for its meaning was to recognize it as a
bullying threat backed with the currencies of jihadist fervor and

invincible ignorance. Like the prophet Isaiah, who beheld the foul sewer of the earth "polluted under the inhabitants thereof," the latter-day bringers of joy and righteousness from the suburbs of Los Angeles and the mountains of Colorado believe themselves obliged to cleanse the world of its impurities—to render justice, reward merit, mete out punishments—and the first few sentences of their joint statement stand as fair indicators of the tone in which they describe the rest of the program:

> We engage in public life because God created our first parents in his image and gave them dominion over the earth. (Genesis 1:27–28). . . . We also engage in public life because Jesus is Lord over every area of life . . . to restrict our stewardship to the private sphere would be to deny an important part of his dominion and to functionally abandon it to the Evil One. To restrict our political concerns to matters that touch only on the private and the domestic spheres is to deny the all-encompassing Lordship of Jesus (Revelation 19:16).

Elsewhere in the document the pastors complain of "the bias of aggressive secularism" so entrenched in the liberal news media that "the presence and role of religion in public life is attacked more fiercely now than ever."

Would that it were so. No citizen can stand for public office in the United States without first pledging allegiance to the King of Kings. Far from being scornful of the messages blown through the trumpets of doom, the news media make a show of their civility and a virtue of their silence; here to please and not to think; every American free to worship the reflection of his or her own fear; no superstition more deserving than another, no imbecile vision in the desert that can't be sold to a talk show, a circus, or the Republican caucus in the House of Representatives.

We used to know better, and to clear away the mess of sanctimony that now seeps into so much of the public mumbling about religion, I find that I'm better served by some of the country's nineteenth-century writers, among them Mark Twain, Ambrose Bierce, and Robert Green Ingersoll, than by a contemporary press too often wrapped in the cellophane of political correctness. Twain discovered in the Bible "noble poetry . . . some clever fables; and some blood-drenched history; . . . a wealth of obscenity; and upwards of a thousand lies"; on the off-chance that a few of his readers had missed the point, he later extended his remarks with a brief sketch of the merciful and Almighty Father revealed in the books of the Old Testament:

> The portrait is substantially that of a man—if one can imagine a man charged and overcharged with evil impulses far beyond the human limit; a personage whom no one, perhaps, would desire to associate with now that Nero and Caligula are dead. In the Old Testament His acts expose His vindictive, unjust, ungenerous, pitiless and vengeful nature constantly. He is always punishing—punishing trifling misdeeds with thousandfold severity; punishing innocent children for the misdeeds of their parents; punishing unoffending populations for the misdeeds of their rulers; even descending to wreak bloody vengeance upon harmless calves and lambs and sheep and bullocks as punishment for inconsequential trespasses committed by their proprietors. It is perhaps the most damnatory biography that exists in print anywhere.

Like Twain, Ingersoll understood that nobody with a sense of humor ever founded a religion, and as the foremost orator of America's Gilded Age, he was famous for the public lectures in which he comforted the sinners and confounded the saints with the tenor of his wit: "Is there an intelligent man or woman now in the world who believes in the Garden of Eden story? If you find any man who believes it, strike his forehead and you will hear an echo. Something is for rent." Or again, on the separation of church and state, "An infinite God ought to be able to protect himself, without going in partnership with State Legislatures."

As an unbaptised child raised in a family that went to church only for weddings and funerals, I didn't encounter the problem of religious belief until I reached Yale College in the 1950s, where I was informed by the liberal arts faculty that it wasn't pressing because God was dead. What remained to be discussed was the autopsy report; apparently there was still some confusion about the cause and time of death, and the undergraduate surveys of Western civilization offered a wide range of options—God disemboweled by Machiavelli in sixteenth-century Florence, assassinated in eighteenth-century Paris by agents of the French Enlightenment, lost at sea in 1834 while on a voyage to the Galápagos Islands, blown to pieces by German artillery at Verdun, garroted by Friedrich Nietzsche on a Swiss Alp, and the body laid to rest in the consulting rooms of Sigmund Freud.

On the evidence presented in the history books, the exit strategy wasn't as important as the good news that the Great Man was well and truly gone. Over a span of nearly 2,000 years, He had let loose upon the earth a sea of blood almost of a match with Lake Superior, and in the long list of religious wars, inquisitions, jousts, massacres, persecutions, and burnings at the stake, I remember the Albigensian Crusade as the baseline measure for all the rest. Pope Innocent III gave his blessing to the program of systematic terror sustained for nearly twenty years against the townspeople of the Languedoc, and to the command of the papal armies he assigned Arnaud-Amalric, the ruling abbot of the Cistercians. When the abbot's troops burned the city of Beziers in 1209 and made prisoners of its 15,000 inhabitants, they asked the supreme monk how they were to distinguish between those still faithful to Holy Church and those who had strayed into the paths of wickedness. "Kill them all!" Amalric is reported to have said. "God will recognize His own." The word to the wise has come down to us through the centuries in the form of policy initiatives blessed by Lucretia Borgia, Torquemada, the bishops of Siena and Rouen, Suleiman the Magnificent, Vlad the Impaler, Generals Erich Ludendorff and Alexander Haig, Josef Stalin, Adolf Hitler, Al Capone, Osama bin Laden, and the U.S. Air Force.

My unassigned wanderings in the Yale libraries and bookstores invariably led to authors with a sense of irony or humor, and by the time I left college in 1956, I assumed that God's once awful and majestic presence had been contained within the walls of a museum or the music of J. S. Bach. The mistake was an easy one to make for a young newspaperman loose in the city of New York in the 1960s with a secular habit of mind and enough money to pay the tithes both to Eros and to Mammon. My travels seldom took me anywhere except to California, and although I heard rumors of the religious enthusiasms roaming the American plains, I chose to regard them as preposterous. If in Florida I sometimes ran across a true believer in an airport or hotel bar, I avoided the embarrassment of a conversation about the Second Coming in much the same way that I'd learned to withhold comment when asked for an opinion by the author of a misshapen novel.

By the middle 1980s, I understood that God had worked another of his miracles, risen from the graves of skepticism and science, moving east from Oklahoma with a great host of gospel-singing Baptists. He began to appear at political rallies clothed in the raiment of Jesus, introduced by the apostles of the newly awakened Christian right as the man to see about buying real estate in heaven—no money down, no homosexuals in the golf shop, every condo equipped with a barbecue pit for the roasting of chestnuts and secular humanists.

As a sales promotion for the sweet hereafter, the message brought by Jerry Falwell's Moral Majority borrowed more heavily from the vicious prophecies of the Old Testament than from the gentler teachings of the New, but during the decade of the 1990s, it attracted increasingly large numbers of increasingly enraged and paranoid disciples who came together as a political constituency in time to provide George W. Bush with a winning margin of electoral votes in last year's presidential election. Responsive to the kind of people on whom it depends for support, the White House grounded the campaign on the twin pillars of fear and intimidation—the promise of never-ending Holy War on terrorism accompanied by political favors for those of the nation's pastors who threatened their congregations with the news that a vote for John Kerry was a one-way ticket to eternal damnation. On the day after the election, Bush received a note from Bob Jones III, president of the eponymously named university in South Carolina: ". . . if you have weaklings around you who do not share your biblical values, shed yourself of them. . . ."

Advice administered as threat conforms to the ethic of the government currently in office in Washington, consistent not only with the character of the deity portrayed in the Old Testament but also with the modus operandi of the Sicilian mafia. Profess loyalty, show respect, launder the money, or expect to wind up whacked or left behind. The born-again capos and underbosses of the Bush Administration (the President himself; Tom DeLay, majority leader in the House; Senators Rick Santorum [R., Pa.] and Sam Brownback [R., Kans.]; Secretary of State Condoleezza Rice) make their bones by robbing the poor to pay the rich and holding fast to the doctrine of preemptive strike, as certain as the prophet Ezekiel that on the day of wrath when the Lord redeems mankind in a flood of purifying fire and a wonder of Hollywood explosions, the faithful and the pure in heart shall find their way home to Paradise.

The guarantee of terrible punishment for God's enemies, combined with the assurance of an ending both happy and profitable for God's business associates, provides the plot for the Left Behind series of neo-Christian fables (thirteen volumes, 62 million copies sold) that have risen in popularity over the last ten years in concert with the spread of fundamentalist religious beliefs and the resurrection of the militant Christ. The co-authors of the books, Tim La-Haye and Jerry P. Jenkins, tell the story of the Rapture on that marvelous and forthcoming day when the saved shall be lifted suddenly to heaven and the damned shall writhe in pain; like most of the prophets who have preceded them to the corporate skyboxes of boundless grace, they express their love of God by rejoicing in their hatred of man. Just as the Old Testament devotes many finely wrought verses to the extermination of the Midianites (also to the butchering of all the people and fatted calves in Moab), LaHaye and Jenkins give upward of eighty pages to the wholesale slaughter of apostates in Boston and Los Angeles, the words as fondly chosen as the film footage in Mel Gibson's *The Passion of the Christ* or the instruments of torture in a seventeenth-century Catholic prison. The twelfth book in the series delights in the spectacle of divine retribution at the battle of Armageddon: "Their innards and entrails gushed to the desert floor, and as those around them turned to run, they too were slain, their blood pooling and rising in the unforgiving brightness of the glory of Christ."

Twenty years ago I would have discounted the stories as childish entertainments comparable to a Tom Clancy or a Harry Potter novel, but the same stupidity now shows up in the "biblically balanced agenda" brought down by the evangelical pastors from Mount Ararat in Colorado and by the gospels of fear and hate espoused by Dr. James Dobson's Focus on the Family. Guided by God's command to impose His sovereignty over "every area of life," public and political as well as private and domestic, Pastor Dobson's apparat endorses political candidates who favor the execution of homosexuals and of doctors who provide abortions. I don't think they're joking. The House of Representatives now shelters 130 members who believe themselves born again in Christ, and in late March, under pressure from the communities of religious fervor gathered in the country's prayer tents, it voted (on behalf of a Florida woman's divine right to life) to replace the laws of the United States with what it was pleased to acknowledge as the will of God.

The faith-based initiative descends upon the multitude in the glorious cloud of unknowing that over the last twenty years has engulfed vast tracts of the American mind in the fogs of superstition—the regressions apparent on the liberal as well as on the conservative aisles of the political argument, evident in the challenges to the teaching of evolution mounted in forty-three states, attested to by the popular belief that Saddam Hussein possessed a magical store of nuclear weapons, by the drainings of public money from the research sciences and the study of history, most wonderfully of all by President Bush's offering his ignorance as the proof of his virtue, claiming that America can rule and govern a world about which it chooses to know as little as possible.

The delusional is no longer marginal, and we err on the side of folly if we continue to grant the boon of tolerance to people who mean to do us harm in the conviction that they receive from Genesis the command "to take dominion over the earth," to build the Kingdom of God, to create the Christian Nation. The proposition is as murderous as it is absurd, and by way of rebuttal we would do well to refer to the sarcasms of Twain and to the intelligence of Ingersoll's essay "God and the Constitution":

When the theologian governed the world, it was covered with huts and hovels for the many, palaces and cathedrals for the few. . . . The poor were clad in rags and skins—they devoured crusts, and gnawed bones. The day of Science dawned, and . . . There is more of value in the brain of an average man of today—of a master-mechanic, of a chemist, of a naturalist, of an inventor, than there was in the brain of the world four hundred years ago.

These blessings did not fall from the skies. These benefits did not drop from the outstretched hands of priests. They were not found in cathedrals or behind altars—neither were they searched for with holy candles. They were not discovered by the closed eyes of prayer, nor did they come in answer to superstitious supplication. They are the children of freedom, the gifts of reason, observation and experience—and for them all, man is indebted to man.

Amen.

May 2005

Condottieri

*All who live in tyranny and hopelessness can know: The United States will not
ignore your oppression or excuse your oppressors. When you stand for your liberty,
we will stand with you.*
—President George W. Bush

Whoever wrote the hero's boast into the President's second
Inaugural Address at least had sense enough to omit the
antecedents for the pronouns. Why spoil the effect by naming
protagonists who might or might not show up for the medal cer-
emonies on the White House lawn? The precaution has proved
the better part of valor during the months subsequent to the January speech, the "you" being seen to refer to a quorum of potentially oil-rich politicians in Iraq, the "we" to the infantry squad
sent with Tom Hanks to save Private Ryan. President Bush looks
to Hollywood for his lessons in geopolitics, and apparently he
likes to think of himself as a military commander in the romantic tradition of Generals George A. Custer and George S. Patton.
His adjutants find it hard to say anything in his presence that
doesn't go well with the sound of bugles, but before he declares
war on Mexico somebody ought to tell him that the American

Army is best equipped to stand and serve not as an invincible combat force but as the world's largest and most heavily armed day-care center.

The several degrees of separation between the mission and the presidential mission statement furnished the national news media in February, March, and April with a steady supply of headlines from sources both foreign and domestic. No lack of "tyranny and hopelessness" abroad, almost all of it excused or ignored because where was the profit to be gained or the glory to be won by standing up for liberty in China, North Korea, Chechnya, Israel, Zimbabwe, or the Sudan? Meanwhile, at home, no end of reports about the scarcity of volunteers eager to play the game of capture the flag in the Atlas Mountains or the Khyber Pass. The latter set of dispatches brought word of the rewards currently being offered to the prospective boots on the ground—bonuses of $90,000 over three years ($20,000 in cash, $70,000 in supplemental benefits), forgiveness of college loans, the promise of citizenship to foreign nationals (currently estimated at 3 percent of the American Army), the acceptance of older recruits (now eligible to the age of thirty-nine), a general lowering of the intellectual and physical requirements (waivers granted for poor test scores, for chronic illness, in some instances for the disability of a criminal record), the chance of a generous pension, an opportunity to study the art of restaurant management.

And yet, despite the inducements and the Army's annual $300 million appropriation for a seductive advertising campaign, the ranks continue to dwindle and thin. Generals speak of "exhausted," "degenerating," "broken" force levels. Recruiting officers give way to unmanly bouts of depression when they fail to enlist more than one soldier for every 120 prospects to whom they show the promotional brochures; so do the reserve units returned, on short notice and without explanation, to another year directing traffic in Iraq and Afghanistan. The desertion rate now stands at 3.1 percent of the active service inductions; of the new recruits coming into camp, 30 percent depart within six months of their arrival.

The flow of dispiriting news provided the Harry Frank Guggenheim Foundation with a casus belli for a seminar staged in New York during the first week of April under the banner "Bearing Arms: Who Should Serve?" As a director of the foundation, I had the chance over the course of two days to hear the question answered by a number of people as strongly opinionated as they were well informed, among them Victor Davis Hanson, the military historian, Charles Moskos, the Northwestern University sociologist and adviser to the Pentagon, Josiah Bunting, president of the Foundation and formerly superintendent of the Virginia Military Institute, U.S. Congressman Charles Rangel (D., N.Y.), who in 2003 proposed legislation to reestablish the military draft. The conversations encompassed a broad range of ancillary topics—America's military history, weapons both ancient and modern, the changes brought about by the enlistment of women in the armed services—but the questions that supplied the energy to the discussion were the ones touching on the reluctance of the country's privileged and well-educated youth (for the most part presumed lost in the desert of materialism) to go to war. Why had the Princeton class of 1956 sent 400 of its 900 graduates to the Army, the class of 2004 only 9 of 1,100? What had become of the homegrown courage that went ashore with "the greatest generation" on the beaches of Normandy and Iwo Jima? Where else if not in the Army was it possible to "share the burden of citizenship," learn the meaning of democracy, find a cure for the disease of selfishness that rots the country's soul?

Although admiring of the reflections derived from the works of Teddy Roosevelt ("Aggressive fighting for the right is the noblest sport the world affords"), I was struck by the fact that nobody took the trouble to consider the nature of the army that young Americans are now being invited to join. What is its purpose, and at whose pleasure does it serve? Judging by the uses to which the all-volunteer army has been put since it was formed in 1973, the defense of the United States ranks very low on its list of priorities. So does the business of waging foreign wars. The domestic political response to the high number of American casualties in Vietnam (57,000 killed, 153,000 wounded) forced the Pentagon to the discovery that it was best to leave the world's noblest sport to well-trained machines and randomly chosen civilians. Although paid for with public money, the Army now operates for the benefit of a primarily private interest, distributing expensive gifts to venal defense contractors, rounding up goons for the oil companies doing merger and acquisition deals in hostile environments, functioning as a prop in presidential-election campaigns, managing a large-scale public-works project that finds employment for the unemployable. The privatization of what was once a public service undoubtedly adds to the country's prestige as well as to the net worth of the consortiums that build planes that don't fly and tanks that sink in the sand, but it cannot be said to constitute a noble cause for which young Americans of any social-economic class—rich, poor, privileged, underachieving—sally gladly forth to fight and die.

The comparison to a day-care center serves both as metaphor and as statement of simple fact. The estimated cost of the life-long health benefits owing to retired veterans and their families over the next ten years now comes up to the sum of $150 billion, which exceeds by $50 billion the Pentagon's annual expenditure on the design of new weapons and the purchase of live ammunition. One of the panelists at the foundation's seminar told of his recent tour of an aircraft carrier in company with its chief medical officer, who pointed out the many and improbable map coordinates at which the ship's crew schedules the hasty assignations of Mars with Venus. He came away from the briefing with the impression of a floating rabbit hutch. Another of the panelists reported 40 percent of our enlisted personnel married to fellow soldiers, an arrangement he thought favorable to women otherwise at a loss to secure the health and education of their children. The recruiting posters and television commercials embody the strength and spirit of the Army as a young man outward bound in a blaze of bravery; the two figures that have come to symbolize the war in Iraq are Lynndie England and Jessica Lynch, both looking not for a way into the halls of military glory but for a way out of the hollows of Appalachian poverty.

As a matter of historical record, the experience of the two young women from West Virginia is more nearly representative of the American attitude toward war than the handsome schoolbook illustrations of Andrew Jackson directing the Battle of New Orleans, Robert E. Lee astride his horse at Gettysburg. Contrary to the claims of the stern moralists who hurl sandbags of furious commentary into the pages of *The Wall Street Journal* and *National Review*, Americans never have had much liking for the heroics cherished by the ancient Romans. Given a good or necessary reason to deploy the military virtues of courage and self-sacrifice, we can rise to the occasion at Bastogne or Guadalcanal, but as a general rule we don't poke around in the cannon's mouth for the Easter eggs of fame and fortune, and if given any choice in the matter, we prefer the civilian virtues—the fast shuffle, the smooth angle, the safe bet.

The shortage of patriots during the Revolutionary War obliged the Continental Army to reward its seasonal help with a 160-acre gift of land; the troops who crossed the Delaware River with George Washington on Christmas Eve, 1776, completed their terms of service on New Year's Day, 1777, refusing to march north to the Battle of Princeton on January 3 until each of them had been paid $10 in gold for another six weeks of labor on the field of honor. President James Madison encountered similar difficulties in July 1814 when a British army arrived in Maryland, intent upon laying waste

to the countryside. Madison issued a requisition for 93,500 militia-
men from what were then the eighteen American states; approximately 6,000 volunteers showed up for the battle of Bladensburg,
where they were promptly dispersed like a flock of birds rising on
the sound of a single gunshot.

The moral of the tale was not lost on John Quincy Adams.
Speaking as the American secretary of state in 1821, he opposed
the sending of the American Navy to liberate the oppressed and
long-suffering peoples of Chile and Colombia from the tyranny of
Spain. Washington that year was seized with delusions of imperial
grandeur not unlike the ones currently walking the White House
dogs; the Spanish viceroys were said to be as cruel and corrupt as
Saddam Hussein, and majorities in both houses of Congress were
eager to carry the flag of freedom south for God, the coffee planta-
tions, and the happiness of all mankind. Adams thought the sentiment fatuous and the policy self-defeating.

America, he said, "goes not abroad in search of monsters to destroy. . . . She would involve herself beyond the power of extrica-
tion, in all the wars of interest and intrigue, of individual avarice,
envy, and ambition, which assume the colors and usurp the standard of freedom. The fundamental maxims of Her Policy would in-
sensibly change from liberty to force."

The sons of liberty were as wary of the Civil War as they had
been careful to avoid enlistment in the Revolutionary War and the
War of 1812. Once it was understood that the march on Richmond
wasn't the holiday jaunt anticipated by the orators north of the Potomac, the federal government was hard-pressed to find soldiers
willing to trample out the vintage where the grapes of wrath were
stored. Between July 1863 and April 1865, the Lincoln Administration sent draft notices to a total of 776,892 men; 161,244 failed
to report, 86,724 paid commutation fees ($3,750 at the current
rate of exchange), 73,607 provided substitutes, 315,509 were examined and ruled exempt; only 46,347 were herded into uniform.

The notion of a citizen army, readily and enthusiastically assembled under the flags of honor, duty, country, emerged from the wreckage of the Second World War. America had been attacked, by Japan at Pearl Harbor and by Germany in the Atlantic Ocean. Peace was not an option, and the American people didn't need to be reminded by a clucking of newspaper columnists that our enemies possessed weapons of mass destruction and that their objectives were murderous. The sense of a common purpose and a national identity bound together in the nucleus of war sustained the government's demand for 10 million conscripts in the years 1941–1945.

The force structure collapsed under the weight of the lies told to the American people by three American presidents trying to find a decent reason for the expedition to Vietnam. Our victory was declared inoperative in April 1975, and for the next quarter of a century when mustering the roll of the all-volunteer Army, the recruiting officers took pains to liken it to a reality-television show. Not the kind of outfit that takes casualties—a vocational school, a summer camp, a means of self-improvement for young men and women lacking the advantages (a decent education, health care, foreign travel) available to Princeton graduates. It had come to be understood that the Pentagon was in the advertising business, projecting images of supreme power in sufficiently heavy calibers to shock a French intellectual and awe an American president. Nobody on the production staff was supposed to get hurt.

Recent events in Iraq have spoiled the sales pitch, which is why the Bush Administration now seeks to carry on its crusade against all the world's evildoers with an army of robots. In Washington on March 26, an Army spokesman described the wonders of the Future Combat Systems (a.k.a. the "technological bridge" to tomorrowland), equipped, at an initial cost of $145 billion over the life expectancy of the miracle, with radio-controlled cannons, tanks, and mortars so godlike in their power and performance as to require next to nothing in the way of food, armor, water, ammunition, or sexual companionship.

The proposal is not without merit. Certainly it meets all the specifications of a government social-welfare program—no military personnel at risk, the day-care centers refurbished and enhanced, enough money lying around loose to win the heart and mind of every proud American in Congress and the weapons trade.

Even so, and not wishing to cast doubts on anybody's patriotism, I think the country might be better served if the corporations fielded their own private armies. The practice is not without precedent. The merchant princes of the Italian Renaissance had as much of a talent for collecting barbarous soldiers as they did for commissioning noble works of art and architecture. If in Michelangelo and Botticelli they could appreciate the presence of genius, so also in the condottieri under the command of Muzio Attendolo and Sigismondo Malatesta they could recognize the high quality of men "insensible to the fear of God," who knew how to "set places on fire, to rob churches . . . imprison priests." Citicorp and ExxonMobil would do well to follow in the footsteps of the Medici—the military operations conceived as venture-capital deals, the soldiers promoted to the rank of shareholders and dressed in uniforms bearing the corporate insignia, the print and broadcast rights firmly under the control of the publicists, the loot divided in accordance with the rules governing the award of contracts in the National Football League.

June 2005

Be Prepared

*A frivolous society can acquire dramatic significance
only through what its frivolity destroys.*
—Edith Wharton

I don't know why so many people continue to insist that we're living in a democracy that somehow would have been recognizable to Franklin D. Roosevelt or even to Richard M. Nixon. The belief is bad for their health and mental stability, in no way conducive to the upkeep of a decent credit rating or an appropriate state of personal hygiene. Or so at least it would appear if I interpret correctly what I've seen of the writers who for the last six months have been showing up in the magazine's offices with stories about the perfidy of the Bush Administration. They come forward with so many proofs of whatever crime against liberty or conscience they happen to have in manuscript or in mind (the war in Iraq, the corruption of Wall Street, the ruin of the schools, the nullification of the United States Senate, etc.) that they run the risk, as did the old Greek heroes who gazed upon the face of Medusa, of being changed into blocks of humorless stone. How can they hope to save the world from its afflictions—to stop the massacre of the innocents in Africa and Asia, preserve the Brazilian rain forest, uplift the multitudes in the slums of Las Vegas and Detroit, fight the war on terror in this darkest hour of the country's peril—if they allow themselves to become sullen and depressed, don't pay attention to their choice of adjective or to the grooming of their shoes? How can they expect to write important and financially rewarding books?

Maybe it's still worth the trouble to wonder when the American democracy lost its footing in Hollywood or Washington, but the historical fact is no more open to dispute than the extinction of the Carolina parakeet or the disappearance of Pickett's brigade on the field at Gettysburg. The setback doesn't mean that we must abandon our belief in the powers of the American imagination or the strength of the American spirit. If we have learned nothing else from the events of the last few years, it's the lesson about the facts being less important than the ways in which they're presented and perceived. Where is the joy in vain regret, where the profit to be gained by selling America short? Somewhere the sun is shining, and only a blind man doesn't look for the silver lining when clouds appear in skies of blue.

Why vilify President George W. Bush as an illiterate lout forwarding freight for a rapacious oligarchy when it's just as easy to see in him almost the entire catalog of virtue listed in the Boy Scout Law—helpful, cheerful, courteous, loyal, friendly, obedient, and clean? The man finds himself burdened with undeserved misfortune in Iraq—American soldiers dying for reasons yet to be explained, money draining into the sand, angry and ungrateful Arabs killing polite and well-behaved Arabs, no good news anywhere in sight—and how does our President respond? Does he reproach his former CIA director, George Tenet, for testifying to the existence of fictitious weapons of mass destruction, dismiss Donald Rumsfeld, his secretary of defense, for ordering a military invasion while under the impression that he was playing a video game in a Virginia penny arcade. No sir, not George W. Bush. We have before us a commander-in-chief who remembers the Boy Scout Oath—to do his best, "to help other people at all times," to keep himself "physically strong, mentally awake, and morally straight." Instead of berating Tenet with charges of incompetence or treason, the President decorates him with the Medal of Freedom; instead of sending Rumsfeld to a rest home among retired admirals in San Diego, he outfits him with another $450 billion to recruit an army of magical machines and pursue America's war on an unknown enemy and an abstract noun.

On a similarly admirable premise (that there is no loyal American undeserving of the nation's trust) the President nominates a bigot as America's ambassador to the United Nations, appoints to the federal appeals courts a synod of judges eager to swear their oaths on the book of Revelation. The decisions speak to the resolve of a man unwilling to take no for an answer, who holds fast both to the Christian theory of redemption and to the American belief in the second chance, the fresh start, and the new beginning—the very same set of doctrines that provided Mark Twain with the plot for *Huckleberry Finn* and furnished the pilgrim fathers with the chance to build in the American wilderness a New Jerusalem free from the encumbrances of history and the storage of prior arrest records.

To denounce the Bush Administration's foreign and domestic policies as the products of false advertising is to misunderstand the character of the American dream. If we don't recognize the fireworks display in Iraq as a sales promotion for America's "credibility in the world," we do ourselves a disservice and dishonor the memory of the Alamo and P. T. Barnum. The pioneers moving west of the Mississippi River in the 1840s chose for their captains men who could hearten them with descriptions of the land of milk and honey awaiting their arrival, like tomorrow's economic recovery or the democratic transformation of Iraq, just around the next bend in the river or across the next range of hills. The caravans had little use for doubts or maps. What was wanted was a leader on the order of Ronald Reagan, a booster blessed with the gift for what Harvard Business School professors define as "aspirational rhetoric," a preacher who could sell the story with or for a song.

As often happened when the joint venture came to grief, the company relied for its salvation on the invisible hand of the divine providence that since the days of James Fenimore Cooper's *Deerslayer* has supplied American expeditions with ways out of traps that ninety-nine times in a hundred would have doomed a lesser breed of men to certain bankruptcy or death—Major Reno and Captain Benteen departing in a timely fashion from the confusion at the Little Bighorn, John D. Rockefeller emerging unscathed from the wreckage of the Standard Oil Trust, President Clinton wriggling free from Monica Lewinsky's thong. We can mention in the same dispatches the resourcefulness of the executives at United Airlines, who as recently as this May solved the riddle of the corporation's insolvency by defaulting on its obligation to pay the $6.6 billion in pensions owed to the 120,000 beneficiaries. So shrewd a stroke of initiative (bold, quick-witted, entrepreneurial) could have occurred only to people well-versed in the practice of rugged individualism.

When it was announced in April that each of the three principal executives at Viacom had awarded himself a total compensation of between $52 million and $56 million for services rendered to a company that in the same year lost $17.5 billion, various small-minded critics in the business press attributed the gesture to the sin of avarice. They missed the better meaning and the higher truth. As is well-known from the teachings of the Bible and Frank Capra movies, a man's spiritual worth cannot be measured by his worldly success. If the executives had sacrificed their principles on the altar of Mammon (bowing to the cynical and vulgar wisdom that looks for some sort of a correlation between a CEO's pay scale and the company's share price), how could they be certain of their inner righteousness? And without a proper sense of self-esteem, how could they muster the courage to lead their investors into the mountains of limitless wealth? Had Brigham Young been cowed by the conventional wisdom harrying the Mormon faithful in the streets of Nauvoo, Illinois, his wagon trains never would have come safely home to the kingdom of Deseret on the shores of the great Salt Lake; if General George S. Patton had listened to the voices of bureaucratic caution, the Third Army never would have reached the bridge at Remagen.

As intrepid as the executives at Viacom but more innovative in his grasp of virtual reality than the accountants at Enron, Bernard Ebbers, the former chief executive of WorldCom, conceived an $11 billion swindle by thinking as far outside the box as did the Wright brothers and Thomas Edison. Fearless on the threshold of the unknown, he ventured over the horizon into the kingdom of imaginary numbers and found gold where none was known to exist. At least 20,000 people lost both their jobs and their pensions, but how else did America settle the Great Plains if not in the company of dead or dying Indians?

Nor did Ebbers flinch in the face of adversity when brought into court last February on charges of criminal fraud. As forthright as the young George Washington contemplating the stump of the chopped-down cherry tree, Ebbers made no attempt to hide behind the screen of a lawyer's weaseling lies. "I don't know about technology" he said, "and I don't know about finance and accounting."

A similar telling of humble, frontier truths distinguished the testimony of L. Dennis Kozlowski, former chief executive of Tyco

International, when two months later he followed Ebbers into a courtroom to explain why he had stolen $150 million from his own company and how he had drawn an illegal profit from the sale of falsely inflated Tyco stock worth $575 million. Another of the trend-setting pioneers in the wilderness of postmodern capitalism, Kozlowski described himself as a visionary misled by charlatans, helpless in the hands of scoundrels. Asked why he neglected to inform the IRS of a $25 million addition to his income in 1999, Kozlowski said, "I was not thinking when I signed my tax return." Asked whether he had such a thing as a computer in his office, he said, "Yes, but just for show."

Look for the silver lining in the confessions of Ebbers and Kozlowski, and instead of a sorry farce the careful reader finds inspirational narrative fit for *Oprah* or a White House press release. Two poor but honest youths born on the wrong side of the tracks rising in the world on the rafts of pluck and grit into the bright light of big-time money and success and there adopted as pets by a news media that praises their brilliance and exclaims over the size of their net worth. Less steadfast individuals might have become confused in the glare of cheap publicity, given way to a sense of social inferiority in the company of even richer corporate folks who kept expensive mistresses, belonged to exclusive clubs, owned zeppelins and 120-foot yachts. Not Ebbers and Kozlowski. As firm of purpose as two Boy Scouts earning merit badges, they acquired more companies, bought bigger yachts, paid $500,000 for wallpaper strewn with hand-painted birds. When neither frog changed into a prince, they didn't lose heart, didn't abandon the pilgrim road to self-affirmation. Ebbers remained in his Mississippi workshop, continuing to manufacture his ever-more-wonderful numbers, homespun and hand-sewn; Kozlowski sallied boldly forth into the high-end consumer markets, collecting an $18 million apartment on Fifth Avenue, and presenting his wife with a party in Sardinia that cost $2 million in company money and delighted the assembled guests with an ice sculpture of Michelangelo's *David* that urinated vodka.

As to the uses of Kozlowski's computer, what else is every political campaign speech and real estate prospectus if not a gesture made "just for show"? It is our glory as a people to prefer the word to the deed, the image of the thing in place of the thing itself. The Department of Homeland Security in early May disclosed that among the antiterrorism devices it had acquired at a cost of $4.5 billion since September 11, 2001, few have proved effective. The radiation monitors at ship terminals can't differentiate between radiation emitted by a nuclear bomb and radiation seeping out of cat litter, ceramic tile, or a crate of bananas. The metal detectors at airports can't be trusted to notice a handgun.

Fortunately for the hope of the country's always better and brighter future, we're blessed with a mainstream news media that shuns sarcasm and knows where to look on the sunny side of life for the silver threads among the gray. A particularly heart-warming proof of the all-volunteer attitude appeared in late May in the pages of *Newsweek*. On May 9 the magazine had published a brief bulletin, anonymously sourced but government-sponsored, to the effect that interrogators in the American prison at Guantánamo Bay, Cuba, had flushed down a toilet a copy of the Koran. The report provoked riots in Pakistan and Afghanistan that killed at least seventeen people, and by May 18, *Newsweek* was accepting the blame for any damage that might have been done to America's good name and reputation in the world.

Although the item had been shown to Pentagon officials prior to its publication, the magazine hadn't apparently received the government's explicit permission to print what it had been told was the truth. Did the editors stand on the right to free expression guaranteed by the First Amendment? Did they observe that it was the American military occupation of Iraq, not *Newsweek*, that was stirring up trouble among the Afghans and Pakistanis? No sir, never in life. The editors followed orders, fell on the grenade, took one for the team. Which is the proper way to behave in a make-believe democracy—show the flag, blow the bugle, learn to see what isn't there.

July 2005

Civil Obedience

After the uprising of the 17th of June
The Secretary of the Writers' Union
Had leaflets distributed in the Stalinallee
Stating that the people
Had forfeited the confidence of the government
And could win it back only
By redoubled efforts. Would it not be easier
In that case for the government
To dissolve the people
And elect another?

—Bertolt Brecht

J ustice Sandra Day O'Connor on the morning of July 1 announced
her retirement from the Supreme Court, and within a matter of
hours the several armies of the imperial right—conservative and
neoconservative, libertarian and evangelical—were on the move to
what their general staff officers perceived as the field of Armageddon. Here at last was the final battle in America's thirty-year culture war, the heaven-sent chance to restructure the Supreme Court
as an office of the Holy Inquisition, to redeem the multitudes who
had forfeited the confidence of Jerry Falwell and the credit-card
companies, to cleanse the country of its sins (among them the
abomination of gay marriage and the blasphemy of legal abortion),
to reinterpret the Constitution as an initial public offering or a program of religious instruction.

The newspapers over the Independence Day weekend reported a

furious ringing of phones in the headquarters tents of the ideologically pure in heart determined to take from the American people as
much of their independence as could be disposed of in the lime pits
of judicial review. Different division commanders were intent upon
different tactical maneuvers (guaranteeing the sanctity of property
rights, assuring the display of the Ten Commandments in the nation's police stations and public schools, bringing the scales of justice into balance with the books of Deuteronomy and Ayn Rand),
but the map arrows all pointed at the same strategic objective. Justice O'Connor's departure offered the first chance in eleven years for
a new appointment to the Supreme Court, and if the vacancy was
soon to be doubled by the ill health and probable retirement of
Chief Justice William Rehnquist, then now was the moment to win
a great and glorious victory for God, mother, the free market, and
the flag.

Operatives at The Federalist Society and The Heritage Foundation bent to the task of searching the Internet for slander with
which to discredit Democratic politicians too well acquainted with
the works of Karl Marx and John Stuart Mill; Tony Perkins, president of the Family Research Council, promised the assistance of
20,000 churches pledged to the quest for a Supreme Court nominee loyal to the judicial philosophy of Justices Antonin Scalia and
Clarence Thomas. The Committee for Justice, formed three years
ago at the behest of Karl Rove and captained by C. Boyden Gray,
formerly White House counsel to the first President George Bush,
undertook to raise the money for a high-definition advertising campaign promoting the appointment of a jurist capable of reading the
Constitution as it was written by the framers—i.e., by a man or
woman miraculously preserved in the placenta of the late eighteenth
century. In time for the Washington television talk shows on Sunday,
July 3, the ranks of conservative seraphim and libertarian cherubim
had been joined by the National Association of Manufacturers
(which assigned a committee of corporation lawyers to review a

nominee's business qualifications), by Paul Weyrich's Free Congress Foundation, by the Ethics and Religious Liberty Commission of the Southern Baptist Convention, by Focus on the Family, whose president, Dr. James C. Dobson, anticipated "a watershed moment in American history" and looked forward to the confrontation coming with a vengeance.

All the allied factions foresaw a rancorous political argument extended through the rest of the summer into the confirmation hearings before the Senate Judiciary Committee in the early autumn, and they took heart from what they regarded, not unreasonably, as the weakness of the opposition. The caucuses of leftist and liberal opinion (pro-abortion and pro-labor, in favor of protecting the environment and the country's dwindling inventory of civil rights) also devoted the Fourth of July weekend to a mustering of forces—staffing phone banks, hiring lobbyists, buying television time, setting up defensive countermeasures, soliciting the support of George Soros and Sean Penn—but the email alerts didn't carry an equivalent weight of passionate intensity and rhetorical armament; nor did they attract as much notice in either the mainline press or the blogosphere. Captain C. B. Gray signed a contract with Fox News to provide commentary tinged with the romance of pious polemic; Ralph G. Neas, president of People for the American Way and the chief organizer of the secular regiments, issued a statement distinguished by its banality—"No matter what side you're on, everything you've believed in, everything you've cared about, everything you've fought for is at stake." The prominent Democrats in Washington (among them Senators Edward M. Kennedy, Harry Reid, and Barbara Boxer) were careful to avoid strong language infected with partisan fervor. They declared their willingness to raise no loud objections if the President contented himself with a "mainstream conservative" in whom they could see trace elements of a mind not totally submerged in an ideological swamp.

The condition appeared to have been met when Mr. Bush named as his Supreme Court appointee John G. Roberts of the United States Court of Appeals for the District of Columbia Circuit. At a White House press conference on the evening of July 19, Judge Roberts was introduced as a non-dogmatic upholder of the Constitution who didn't confuse the moral with the civil law, and over the course of the next several days what could be learned of his record and character tended to confirm that supposition—fifty years old, born in Buffalo, New York, and educated at Harvard Law School, a devout Roman Catholic, by all accounts an amiable and remarkably intelligent man, conservative in his politics but by no means unreasonable in judgment or rabid in termperament. Because he had served on the Appellate bench for only two years, he hadn't had occasion to provide enough written opinions from which to draw inferences about his judicial philosophy. On the evidence of the judge's prior service as deputy solicitor general in the first Bush Administration, it could be assumed that he disapproved of abortion and favored the enlargement of the federal police power, but the lack of specific information temporarily disarmed his prospective enemies entrenched in the salients on the liberal left.

Nor were there any public expressions of disappointment from the cadres of the militant right. Off the record and not for attribution, the avenging angels of the spiritual counterreformation presumably continued to hope and plan for a permanent solution to the problem of a democracy too loose in its morals and too careless of its freedoms; on the record and when smiling for the cameras, they praised Judge Roberts as a man likely to interpret the Constitution as an instrument designed not to grant or create rights but to take them away, and they were glad to count him as an ally in the culture war that for the last thirty years has taken as its cause the "moral anarchy" sprung from the Pandora's box of the 1960s.

The nation's youth during that too noisy and fanciful a decade made a disgraceful spectacle of their poor conduct and unseemly deportment. Godless liberals poisoned the wells of learning at the Ivy League universities; feminists seized the radio station in Berkeley, California, and destroyed the Bing Crosby records; homosexuals infiltrated the U.S. Navy and the Hollywood movie studios; black people stopped eating watermelon; pornography flourished; culture died. Everything that has since gone wrong for the country (the loss of the Vietnam War, the reversals of fortune in the stock market, the national addiction to expensive drugs and cheap food) follows not from any fault in the political or economic systems but from the flaws of individual character—entirely too many people unwilling to accept personal responsibility, drifting away from the old Norman Rockwell faith in marriage, thrift, hard work, chastity, ennobling self-sacrifice. The American people therefore stand to be corrected, weaned from the teat of mindless decadence, put through the mill of a moral awakening.

The uplifting program of social hygiene fits the agenda of both the religious and the free-market right, and among all the company of public scolds (William Bennett, George Will, Alan Bloom, Rush Limbaugh, Newt Gingrich, etc.) none has been more strident or excessively praised over the last thirty years than Robert Bork, the federal appellate judge who in 1987 was denied appointment to the Supreme Court. Eighteen years later it comes as no surprise that Bork should appear, in the newspaper op-ed pages and on the Washington talk-show circuit, as the lead soloist in the chorus of voices demanding the complete and total annihilation of modern liberalism and all its works. Preaching his familiar sermon for the *Wall Street Journal*'s editorial page on July 5, Bork raged against "homosexual marriage," "racial and gender discrimination at the expense of white males," the criminal justice system "tipping the balance in favor of criminals," the Supreme Court corrupted by its liberal members, by their "philosophical incompetence" and their "disdain for the historic Constitution. . . ." Interviewed in late June by the *New York Times,* which identified him as "a prolific lecturer and author who functions as a kind of shadow justice" (i.e., America's conscience-in-exile), Bork directed his most caustic insults at Justice Anthony M. Kennedy, the Supreme Court judge regarded by the antidemocratic right as an overly genial believer in tolerance and consensus, and therefore, in the words of Pastor Dobson, "the most dangerous man in America." Bork cited Kennedy's observation that "at the heart of liberty is the right to define one's own concept of existence, of meaning, of the universe, and of the mystery of human life," in order to hold it up to scorn and ridicule. "What the Hell does that mean?" he said.

Bork's mean-mouthed suspicions accord with the rightward shifts of American thought and opinion that since September 2001 have sustained the Bush Administration's backing of its self-serving war on terror with the currency of fear—in line with the proliferation of checkpoints and surveillance cameras, with the medieval wall (twelve feet high, ten miles in circumference, topped up with razor wire) that surrounds the American embassy in Baghdad, with the legislation now before the Congress intended to strengthen the USA Patriot Act with the power of the administrative subpoena. Approved by the President and endorsed by the more punitive Republicans in both the Senate and the House of Representatives, the new and improved investigative technology allows the FBI to access at will any individual's financial, medical, and employment records without that person's knowledge, without the prior approval of a judge or a grand jury.

At will, as a disciplinary measure, intended (as per the suggestion in Bertolt Brecht's poem "The Solution") "to dissolve the people and elect another." Brecht wrote the poem in response to the disruptions in East Germany on June 17, 1953, when 300,000 industrial workers in 272 towns and cities staged a general strike and by so doing showed themselves unworthy of the government's respect. East Germany at the time lacked the resources to act on Brecht's advice. The Albanians presumably were available as a substitute people, but the baggage handling presented insuperable logistical difficulties, and so the Communist authorities settled for the instructional apparatus of a police state.

The solution doesn't answer to our own postmodern circumstance; too many Americans have too much money (enough to buy their own virtual-reality editions of both Heaven and Hell), and if we took seriously the complaints about the flaunting of pernicious liberties, we might have to close down the gambling casinos as well as the New York fashion shows and the traffic in prescription drugs. Much better to teach the lessons of obedience with heartwarming examples of proper conduct and correct behavior, to gradually disappear people who don't come up to the factory standard of the model citizen. Certainly we've begun to take important steps in the right direction. The news and entertainment media don't spoil the decoration of their show windows with unsightly forms of physical or intellectual depravity (no left-wing university professors, no slumdwellers, few women or horses without good teeth); the Congress denies to the poorer grades of domestic help the indulgence of adequate health care and a decent education, thus ensuring their continued invisibility and maybe, with luck and another three or four foreign wars, their eventual extinction.

Let the Supreme Court fall safely into the hands of magistrates like those known to the Puritans in seventeenth-century Massachusetts as inspectors of souls, and within another ten or fifteen years we might manage to cultivate a new breed of true American— citizens of tomorrow, proud of their deodorants and bathroom floors, quick to report a neighbor seen talking to an Arab or smoking a cigarette, so accustomed to the lack of privacy that they dare not read Mark Twain's *Huckleberry Finn,* much less eat a peach. At this still early stage of metamorphosis it's probably too soon to hope for a population of deaf-mutes, but I can imagine every fully realized twenty-first-century American to be as innocent of politics as a sun-dried tomato or a quart of milk, grateful for the occasional permissions to make love to one's lawful spouse (careful to remember while doing so that Jesus is in one's heart), dreaming of a career in law enforcement.

Under the rules of procedure governing the Senate Judiciary Committee's autumn confirmation hearings, Judge Roberts will not be required to touch upon what the first President Bush was wont to call "the vision thing." When asked for his views on abortion, property rights, civil liberties, or the mating of stem cells, he can refuse to answer on the ground that the issue might one day come before the Court. Similar to the gifts of silence granted to criminals under the protection of the Fifth Amendment, the exercise of judicial privilege will guarantee the absence of either free or coherent speech. If the hearings fail to provoke an argument about the salvation of the nation's soul sufficiently venomous to merit the blessing of the Prophet Isaiah, the apostles of the radical and reactionary right can always relieve their frustrations by the staging of religious festivals on Capitol Hill—medieval jugglers throwing balls for the Virgin Mary, Christian morality plays and performing bears, excited women holding aloft the papier-mâché head of the sainted Bork, crowns of thorns and flights of doves, Mel Gibson staggering up Pennsylvania Avenue under the weight of a blood-soaked cross.

September 2005

On Message

But I venture the challenging statement that if American democracy ceases to move forward as a living force, seeking day and night by peaceful means to better the lot of our citizens, then Fascism and Communism, aided, unconsciously perhaps, by old-line Tory Republicanism, will grow in strength in our land.
—Franklin D. Roosevelt, November 4, 1938

*I*n 1938 the word "fascism" hadn't yet been transferred into an abridged metaphor for all the world's unspeakable evil and monstrous crime, and on coming across President Roosevelt's prescient remark in one of Umberto Eco's essays, I could read it as prose instead of poetry—a reference not to the Four Horsemen of the Apocalypse or the pit of Hell but to the political theories that regard individual citizens as the property of the government, happy villagers glad to wave the flags and wage the wars, grateful for the good fortune that placed them in the care of a sublime leader. Or, more emphatically, as Benito Mussolini liked to say, "Everything in the state. Nothing outside the state. Nothing against the state."

The theories were popular in Europe in the 1930s (cheering crowds, rousing band music, splendid military uniforms), and in the United States they numbered among their admirers a good many important people who believed that a somewhat modified form of fascism (power vested in the banks and business corporations instead of with the army) would lead the country out of the wilderness of the Great Depression—put an end to the Pennsylvania labor troubles, silence the voices of socialist heresy and democratic dissent.

Roosevelt appreciated the extent of fascism's popularity at the political box office; so does Eco, who takes pains in the essay "Ur-Fascism," published in *The New York Review of Books* in 1995, to suggest that it's a mistake to translate fascism into a figure of literary speech. By retrieving from our historical memory only the vivid and familiar images of fascist tyranny (Gestapo firing squads, Soviet labor camps, Mussolini's riding boots), we lose sight of the faith-based initiatives that sustained the tyrant's rise to glory. The several experiments with fascist government, in Russia and Spain as well as in Italy and Germany, didn't depend on a single portfolio of dogma, and so Eco, in search of their common ground, doesn't look for a unifying principle or a standard text. He attempts to describe a way of thinking and a habit of mind, and on sifting through the assortment of fantastic and often contradictory notions—Nazi paganism, Franco's National Catholicism, Mussolini's corporatism, etc.—he finds a set of axioms on which all the fascisms agree. Among the most notable:

The truth is revealed once and only once.

Parliamentary democracy is by definition rotten because it doesn't represent the voice of the people, which is that of the sublime leader.

Doctrine outpoints reason, and science is always suspect.

Critical thought is the province of degenerate intellectuals, who betray the culture and subvert traditional values.

The national identity is provided by the nation's enemies.

Argument is tantamount to treason.

Perpetually at war, the state must govern with the instruments of fear.

Citizens do not act; they play the supporting role of "the people" in the grand opera that is the state.

Eco published his essay ten years ago, when it wasn't as easy as it has since become to see the hallmarks of fascist sentiment in the character of an American government. Roosevelt probably wouldn't have been surprised. Among old-line Tory Republicans, he'd encountered enough opposition to both the New Deal and to his belief in such a thing as a United Nations to judge the force of America's racist passions and the ferocity of its anti-intellectual prejudice. As he may have guessed, so it happened. The American democracy won the battles for Normandy and Iwo Jima, but the victories abroad didn't stem the retreat of democracy at home, after 1968 no longer moving "forward as a living force, seeking day and night to better the lot" of its own citizens. Now that sixty years have passed since the bomb fell on Hiroshima, it doesn't take much talent for reading a cashier's scale at Wal-Mart to know that it is fascism, not democracy, that won the heart and mind of America's "Greatest Generation," added to its weight and strength on America's shining seas and fruited plains.

A few sorehead liberal intellectuals continue to bemoan the fact, write books about the good old days when everybody was in charge of reading his or her own mail. I feel their pain and share their feelings of regret, also wish that Cole Porter was still writing songs, that Jean Harlow and Robert Mitchum hadn't quit making movies. But what's gone is gone, and it serves nobody's purpose to deplore the fact that we're not still riding in a coach to Philadelphia with Thomas Jefferson. The attitude is cowardly and French, symptomatic of effete aesthetes who refuse to change with the times.

*   *   *

As set forth in Eco's list, the fascist terms of political endearment are refreshingly straightforward and mercifully simple, many of them already accepted and understood by a gratifyingly large number of our most forward-thinking fellow citizens, multitasking and safe with Jesus. It does no good to ask the weakling's pointless question, "Is America a fascist state?" We must ask instead, in a major rather than a minor key, "Can we make America the best damned fascist state the world has ever seen," an authoritarian paradise deserving the admiration of the international capital markets, worthy of "a decent respect to the opinions of mankind"? I wish to be the first to say we can. We're Americans; we have the money and the know-how to succeed where Hitler failed, history has favored us with advantages not given to the early pioneers.

WE DON'T HAVE TO BURN ANY BOOKS

The Nazis in the 1930s were forced to waste precious time and money on the inoculation of the German citizenry, too well-educated for its own good, against the infections of impermissible thought. We can count it as a blessing that we don't bear the burden of an educated citizenry. The systematic destruction of the public-school and library systems over the last thirty years, a program wisely carried out under administrations both Republican and Democratic, protects the market for the sale and distribution of the government's propaganda posters. The publishing companies can print as many books as will guarantee their profit (books on any and all subjects, some of them even truthful), but to people who don't know how to read or think, they do as little harm as snowflakes falling on a frozen pond.

WE DON'T HAVE TO DISTURB, TERRORIZE, OR PLUNDER THE BOURGEOISIE

In Communist Russia as well as in Fascist Italy and Nazi Germany, the codes of social hygiene occasionally put the regime to the trouble of smashing department-store windows, beating bank managers to death, inviting opinionated merchants on complimentary tours (all expenses paid, breathtaking scenery) of Siberia. The resorts to violence served as study guides for freethinking businessmen reluctant to give up on the democratic notion that the individual citizen is entitled to an owner's interest in his or her own mind.

The difficulty doesn't arise among people accustomed to regarding themselves as functions of a corporation. Thanks to the diligence of our news media and the structure of our tax laws, our affluent and suburban classes have taken to heart the lesson taught to the aspiring serial killers rising through the ranks at West Point and the Harvard Business School—think what you're told to think, and not only do you get to keep the house in Florida or command of the Pentagon press office but on some sunny prize day not far over the horizon, the compensation committee will hand you a check for $40 million, or President George W. Bush will bestow on you the favor of a nickname as witty as the ones that on good days elevate Karl Rove to the honorific "Boy Genius," on bad days to the disappointed but no less affectionate "Turd Blossom." Who doesn't now know that the corporation is immortal, that it is the corporation that grants the privilege of an identity, confers meaning on one's life, gives the pension, a decent credit rating, and the priority standing in the community? Of course the corporation reserves the right to open one's email, test one's blood, listen to the phone calls, examine one's urine, hold the patent on the copyright to any idea generated on its premises. Why ever should it not? As surely as the loyal fascist knew that it was his duty to serve the state, the true American knows that it is his duty to protect the brand.

Having met many fine people who come up to the corporate mark—on golf courses and commuter trains, tending to their gardens in Fairfield County while cutting back the payrolls in

Michigan and Mexico—I'm proud to say (and I think I speak for all of us here this evening with Senator Clinton and her lovely husband) that we're blessed with a bourgeoisie that will welcome fascism as gladly as it welcomes the rain in April and the sun in June. No need to send for the Gestapo or the NKVD; it will not be necessary to set examples.

WE DON'T HAVE TO GAG THE PRESS OR SEIZE THE RADIO STATIONS

People trained to the corporate style of thought and movement have no further use for free speech, which is corrupting, overly emotional, reckless and ill-informed, not calibrated to the time available for television talk or to the performance standards of a Super Bowl halftime show. It is to our advantage that free speech doesn't meet the criteria of the free market. We don't require the inspirational genius of a Joseph Goebbels; we can rely instead on the dictates of the Nielsen ratings and the camera angles, secure in the knowledge that the major media syndicates run the business on strictly corporatist principles—afraid of anything disruptive or inappropriate, committed to the promulgation of what is responsible, rational, and approved by experts. Their willingness to stay on message is a credit to their professionalism.

The early twentieth-century fascists had to contend with individuals who regarded their freedom of expression as a necessity—the bone and marrow of their existence, how they recognized themselves as human beings. Which was why, if sometimes they refused appointments to the state-run radio stations, they sometimes were found dead on the Italian autostrada or drowned in the Kiel Canal. The authorities looked upon their deaths as forms of self-indulgence. The same attitude governs the agreement reached between labor and management at our leading news organizations. No question that the freedom of speech is extended to every

American—it says so in the Constitution—but the privilege is one that musn't be abused. Understood in a proper and financially rewarding light, freedom of speech is more trouble than it's worth—a luxury comparable to owning a racehorse and likely to bring with it little else except the risk of being made to look ridiculous. People who learn to conduct themselves in a manner respectful of the telephone tap and the surveillance camera have no reason to fear the fist of censorship. By removing the chore of having to think for oneself, one frees up more leisure time to enjoy the convenience of the Internet services that know exactly what one likes to hear and see and wear and eat.

WE DON'T HAVE TO MURDER THE INTELLIGENTSIA

Here again, we find ourselves in luck. The society is so glutted with easy entertainment that no writer or company of writers is troublesome enough to warrant the compliment of an arrest, or even the courtesy of a sharp blow to the head. What passes for the American school of dissent talks exclusively to itself in the pages of obscure journals, across the coffee cups in Berkeley and Park Slope, in half-deserted lecture halls in small Midwestern colleges. The author on the platform or the beach towel can be relied upon to direct his angriest invective at the other members of the academy who failed to drape around the title of his latest book the garland of a rave review.

The blessings bestowed by Providence place America in the front rank of nations addressing the problems of a twenty-first century, certain to require bold geopolitical initiatives and strong ideological solutions. How can it be otherwise? More pressing demands for always scarcer resources; ever larger numbers of people who cannot be controlled except with an increasingly heavy hand of authoritarian guidance. Who better than the Americans to lead the fascist renaissance, set the paradigm, order the preemptive strikes? The existence of mankind hangs in the balance; failure is not an option. Where else but in America can the world find the visionary intelligence to lead it bravely into the future—Donald Rumsfeld our Dante, Turd Blossom our Michelangelo?

I don't say that over the last thirty years we haven't made brave strides forward. By matching Eco's list of fascist commandments against our record of achievement, we can see how well we've begun the new project for the next millennium—the notion of absolute and eternal truth embraced by the evangelical Christians and embodied in the strict constructions of the Constitution; our national identity provided by anonymous Arabs; Darwin's theory of evolution rescinded by the fiat of "intelligent design"; a state of perpetual war and a government administering, in generous and daily doses, the drug of fear; two presidential elections stolen with little or no objection on the part of a complacent populace; the nation's congressional districts gerrymandered to defend the White House for the next fifty years against the intrusion of a liberal-minded president; the news media devoted to the arts of iconography, busily minting images of corporate executives like those of the emperor heroes on the coins of ancient Rome.

An impressive beginning, in line with what the world has come to expect from the innovative Americans, but we can do better. The early twentieth-century fascisms didn't enter their golden age until the proletariat in the countries that gave them birth had been reduced to abject poverty. The music and the marching songs rose with the cry of eagles from the wreckage of the domestic economy. On the evidence of the wonderful work currently being done by the Bush Administration with respect to the trade deficit and the national debt—to say nothing of expanding the markets for global terrorism—I think we can look forward with confidence to character-building bankruptcies, picturesque bread riots, thrilling cavalcades of splendidly costumed motorcycle police.

October 2005

Slum Clearance

The comfort of the rich rests upon an abundance of the poor.
—Voltaire

On Monday, August 29, a category 4 hurricane slammed into New Orleans with winds reaching 140 miles an hour, and by Thursday, September 1, the city looked just about the way a doomed city is supposed to look according to the Book of Revelation. Which, given the faith-based political theory currently in office in Washington, should have surprised nobody. For the last thirty years the scribes and Pharisees allied with the several congregations of both the radical and the reactionary right have been preaching the lesson that government is a sink of iniquity—by definition inefficient, unjust, wasteful, and corrupt, a mess of lies deserving neither the trust nor the affection of true Americans. True Americans place their faith in individual initiative, moral virtue, and personal responsibility, knowing in their hearts that government is the enemy of the people, likely to do more harm than good.

So it proved in New Orleans during the first week of September.

At every level of officialdom—city, parish, state, and federal—the tribunes of the people met the standard of bureaucratic futility and criminal negligence imputed to them by two generations of Republican publicists, and within the few days before, during, and after the hurricane's arrival, they managed to facilitate the loss of life, liberty, and property for several hundred thousand of their fellow citizens. The devastation fell somewhat short of the biblical prophecy—no blood in the sea, the floodwaters unsmitten with the bloom of Wormwood, no angels overhead armed with the trumpets of Woe; even so, despite the absence of giant locusts wearing breastplates of iron, about as satisfactory a result as could be hoped for from a government public-works program—the storm warnings ignored or discredited, the levees in a reliably shoddy state of repair, 1 million people left homeless in the mostly uninsured wreckage scattered across 90,000 square miles in four states, dead animals drifting in the New Orleans sewage and rotting on the beaches of Biloxi, the sick and elderly dying of thirst in the stench and heat of the Superdome, poisonous snakes making the rounds of hospital emergency rooms, rats gnawing at the corpses of the drowned.

Even more impressive than the scale of the calamity was the laissez-faire response of the government officials who understood that it was not their place to question, much less attempt to interfere with, an act of God. When confronted with scenes of anguish that might have tempted overly emotional public servants to ill-considered activisms, the Department of Homeland Security held fast to the policy of principled restraint. Spendthrift liberals rush to help people who refuse to help themselves; prudent conservatives know that such efforts smack of socialism. The residents of New Orleans had been told to evacuate the city before the hurricane came ashore, and if they didn't do so, well, whose fault was that? Government cannot be held responsible for the behavior of people who don't follow instructions, aren't mature enough to carry an American Express card or drive an SUV.

Every now and then, of course, government must show concern for the country's less fortunate citizens—the gesture is deemed polite in societies nominally democratic—and two days after the flooding submerged most of New Orleans under as much as fifteen feet of foul and stagnant water, President George W. Bush graciously cut short his Texas vacation to gaze upon the ruined city from the height of 2,500 feet. Air Force One remained overhead for thirty-five whole minutes, which was long enough to impress upon the President the comparison to a big-budget Hollywood disaster movie. To the White House aides-de-camp aboard the plane he was reported to have said, "It's devastating, it's got to be doubly devastating on the ground." A sensitive observation, indicating that he had noticed something seriously amiss—small houses floating in the water, big boats moored in trees. A president crippled by too active an imagination might have made the mistake of wanting to see for himself the devastation on the ground, possibly even going so far as to say a few words to the evacuees in the Superdome. But the newscasts were loud with rumors of armed gangs of unattractive black people looting convenience stores and raping infant girls, and if one or more of the mobs happened to incite a riot, the liberal news media would publish unpleasant photographs and draw unpatriotic conclusions. Better to wait until the army had set up a secure perimeter.

By Friday, September 2, four days after the hurricane made landfall, enough military units were in place to allow the President to upgrade the demonstration of his concern with the staging of resolute drop-bys in Louisiana, Mississippi, and Alabama. But if it was clear from his manner that he wished to convey sympathy and offer encouragement, it was also clear that he was at a loss to relate the words in the air to the "doubly devastating" death and destruction on the ground. Standing tall in shirtsleeves in front of the cameras in Mobile, he acknowledged the misfortune that had befallen his good friend Senator Trent Lott (R., Miss.): "The good news is, and its hard for some to see it now—that out of this chaos is going to come a fantastic Gulf Coast, like it was before. Out of the rubbles of Trent Lott's house—he's lost his entire house— there's going to be a fantastic house, and I'm looking forward to sitting on the porch." Later that same day, departing from the airport in New Orleans, the President hit the note of solemnly conservative compassion appropriate to an HBO production of the decline and fall of Rome: "You know, I'm going to fly out of here in a minute, but I want you to know that I'm not going to forget what I've seen."

Most of the other government spokespersons within range of a microphone during the first week in September might as well have been relaying their remarks by satellite from a map room in Bermuda. By Thursday, September 1, reports from the scene at the New Orleans Convention Center had been repeatedly broadcast on every network in the country—several thousand people without food or water, all of them desperate, quite a few of them dying. The news hadn't reached Michael Chertoff, director of the Department of Homeland Security in Washington, who had waited a judicious thirty-six hours after the storm's arrival before declaring it "an incident of national significance." To an interviewer from National Public Radio, Chertoff said, "I've not heard a report of thousands of people in the Convention Center who don't have food and water." The people in question presumably hadn't filled out the necessary forms. Nor had they informed Michael Brown, director of the Federal Emergency Management Agency, who also hadn't heard of any trouble at the Convention Center and who told Wolf Blitzer on September 1, "Considering the dire circumstances that we have in New Orleans, virtually a city that has been destroyed, things are going relatively well." Which was the preferred tone of voice throughout the rest of the week on the part of the Washington gentry doing their best to take an interest in people they neither knew not wished to know.

Former First Lady Barbara Bush on September 5, reviewing the condition of the hurricane flood evacuees in the Houston Astrodome: "What I'm hearing, which is sort of scary, is that they all want to stay in Texas. Everybody is so overwhelmed by the hospitality. And so many of the people in the arena here, you know, were underprivileged anyway, so this (chuckle) is working very well for them."

GOP strategist Jack Burkman, September 7: "I understand there are 10,000 people dead. It's terrible. It's tragic. But in a democracy of 300 million people, over years and years and years, these things happen."

September 8, First Lady Laura Bush: "I also want to encourage anybody who was affected by hurricane Corrina [sic] to make sure their children are in school."

House Majority Leader Tom DeLay, September 9, bucking up the spirits of three young hurricane evacuees from New Orleans at the Astrodome: "Now tell me the truth, boys, is this kind of fun?"

Earlier in the week Mrs. Bush might have been pardoned for mistaking the name of the hurricane—hurricanes come and go in the same way that summer disaster movies come and go, and only a bleeding-heart leftist would expect the theatergoers in a Washington screening room to remember which is which—but by September 8 the news reports from New Orleans and points east were indicating an even more feckless government response than previously had been supposed—the USS *Bataan,* fully supplied with medical facilities, held at a safe distance offshore for reasons unexplained, National Guard units delayed in the confusions of bureaucratic move and countermove, the dysfunction of FEMA understood as the result of the nepotistic hiring of its senior management, trucks bringing ice and water rerouted to South Carolina, evacuees herded onto planes without being told where the planes were bound, the order to evacuate New Orleans made impractical by the simultaneous disappearance of the city's public transportation systems.

As it became increasingly evident that the storm had inflicted its heaviest damage on people who were poor, illiterate, and predominantly black, what emerged from the Mississippi mud was the ugly recognition of the United States as a society divided against itself across the frontiers of race and class. Not "one nation under God, indivisible, with liberty and justice for all" but two nations, divisible by bank account, with liberty and justice for those able to pay the going rate for a government pimp.

The unwelcome sight evoked angry shouts of Woe from all the trumpets of the news media—outraged editorials, harsh questions from television anchorpersons ordinarily as mild as milk, a rising tide of bitter reproach from politicians both Democratic and Republican. The abrupt decline in the President's approval ratings prompted his press agents to send him on a frenzied round of image refurbishment—Mr. Bush holding a press conference to accept responsibility for the federal government's storm-related failures, Mr. Bush at the National Cathedral in Washington, declaring a "National Day of Prayer and Remembrance," Mr. Bush back again on the Gulf Coast, posed in front of the stage-lit St. Louis Cathedral in New Orleans, promising to do and spend "what it takes" ($100 billion, maybe $200 billion) to restore "the passionate soul" of the dead city.

If the performances weren't as uplifting as the President might have hoped, the fault possibly was to be found in his inability to hide the fact of his genuine and irritated surprise. What was everybody complaining about, for God's sake? Who didn't know that America was divided into a nation of the rich and a nation of the poor? What else had every self-respecting Republican politician been doing for the last thirty years if not bending his or her best efforts to achieve that very purpose? Didn't anybody remember the words of the immortal Ronald Reagan's first inaugural address: "Government is not the solution to our problem; government is the problem"? Had everybody forgotten the noble question asked and answered in 1987 by Margaret Thatcher, that great and good British prime minister: "Who is society? There is no such thing! There are individual men and women; there are families"? Some families make it to higher ground; others don't. Such is the way of the world and the natural order of things, visible every day in the pictures from Africa on CNN. Why else was the Republican Party so popular—elected to the White House, put in charge of the Congress and the Supreme Court—if not to give to the haves and take from the have-nots? It wasn't as if anybody, least of all President Bush, had made any secret of the project. All the major legislation passed by Congress over the last five years—the transportation bill, the Medicare prescription bill, the tax bills favoring corporations and wealthy individuals, the bankruptcy bill, etc.—strengthens the power of money to limit and control the freedom of individuals. During the early weeks of September, when countless thousands of people on the Gulf Coast were sorely in need of rescue, Senator Bill Frist (R., Tenn.), the Republican majority leader in the Senate, never once lost sight of the more urgent rescue mission, which was to press forward the legislation intended to privatize Social Security and eliminate the estate tax. Senator Frist is a doctor but first and foremost a loyal Republican and a man who knows how to order his priorities—before the hand on the heart, the thumb on the coin.

As surprised as the President by the grumbling noises in the suddenly and uncharacteristically conscience-stricken media, a heavenly host of Republican preachers and politicians was quick to shift the story into the True American context of individual initiative, moral virtue, and personal responsibility. Thus Senator Rick Santorum (R., Pa.): "I mean, you have people who don't heed those warnings and then put people at risk as a result of not heeding those warnings. There may be a need to look at tougher penalties on those who decide to ride it out and understand that there are consequences to not leaving."

Consequences also for not leading one's life in accordance with the instructions given in the Bible, the point made in the seconding of Senator Santorum's motion by numerous spokesmen for Christ. Thus the pastor of the New Covenant Fellowship of New Orleans: "New Orleans now is free of Southern Decadence, the sodomites, the witchcraft workers, false religion—it's free of all these things now." Or again, more subtly, by the Columbia Christians for Life. The organization correlates storm tracks with cities harboring abortion clinics and supplemented its press release referring to the five such establishments in New Orleans with a satellite photograph that "looks like a fetus facing to the left (west) in the womb, in the early weeks of gestation."

Not a natural disaster, the hurricane, but a blessing in disguise, so seen and much appreciated by the forward-thinking parties of enlightened Republicanism. To the readers of the *Wall Street Journal* on September 9, Congressman Richard Baker (R., La.), brought the good news of a divinely inspired slum-clearance project. "We finally cleaned up public housing in New Orleans," he said. "We couldn't do it, but God did."

As is well known and understood in the elevated circles of Republican political thought, God helps those who help themselves, and on September 13 *Time* magazine quoted an unnamed White House source confirming the miracle of the loaves and fishes soon to be visited upon the well-connected servants of the Lord in Louisiana, Mississippi, and Alabama. "Nothing can salve the wounds like money . . . you'll see a much more aggressively engaged President, traveling to the Gulf Coast a lot and sending a lot of people down there."

By the time it comes to writing next month's Notebook, I expect that we'll have had the chance to count the ways in which the master chefs of our indolent but nevertheless ravenous government can carve the body of Christ into the sweetmeats of swindle and the drumsticks of fraud.

November 2005

The Simple Life

Not to know what happened before one was born is always to be a child.
—Cicero

Standing in front of a New Orleans cathedral stage-lit to resemble one of Disney's magic castles, President George W. Bush on the evening of Thursday, September 15, told the country a fairy tale. His administration, he said, would do "what it takes," spare no expense to make good the losses inflicted two weeks earlier by Hurricane Katrina on the good people of Alabama, Mississippi, and Louisiana. The stock market knew what was meant—some of the people, not all of the people; the preferred few as distinct from the hapless many—and in the next Monday's trading, prices moved comfortably up for companies positioned to find in the wreckage the saving grace of a no-bid government contract. Already on September 8, Congress had appropriated $62.3 billion for the relief and reconstruction of the Gulf Coast, the money up and running from the drainpipes on Capitol Hill at the rate of $300 million a day, but here was the President promising an even greater flood of deliverance (for Vice President Dick Cheney's duck-shooting companions at Halliburton if not for the once-upon-a-time residents of Gulfport and Biloxi), and by Wednesday, September 21, every well-placed receptacle in Washington was expecting the eventual cash flow to crest at a high-water mark somewhere in the vicinity of $200 billion.

Agreed as to the amount of silver in the lining of the cloud, the interested parties also were unanimous in their prediction that much of it would underwrite a Category 5 deluge of fraud, graft, corrupt self-dealing, and outright theft as awe-inspiring as the hurricane. Given the character and disposition of the political operatives currently in control of the government in Washington, how or why would it be otherwise? We have a majority leader in the House of Representatives, Tom DeLay (R., Tex.), indicted on September 28 and again on October 3 on felony charges of conspiracy and money-laundering, a majority leader in the Senate, Bill Frist (R., Tenn.), currently under investigation by the Securities and Exchange Commission for possibly crooked dealings in the stock market, the President's senior adviser, Karl Rove, recalled for a fourth time to testify before a federal grand jury about the White House's troubles with the Intelligence Identities Protection Act. Add to the list of boldfaced names the instances of criminal misrule that appear routinely in every morning's newspaper, and it's fair to say that in Washington at the moment we have a government owned and operated by a rapacious oligarchy that seeks to privatize—i.e., appropriate or destroy—the public infrastructure (schools, roads, air, water, power plants, bridges, levees, hospitals, forests, broadcast frequencies, wetlands, birds) that provides the country with the foundations of its common enterprise. In line with the well-established policy of enlightened selfishness, why not therefore privatize the salvage of the Gulf Coast, look upon the venture as a game of faro aboard a Mississippi gambling boat in which the odds favor the gentlemen at the table with the marked cards and the stacked deck? As early as the first week in October the house rules were clearly posted on the front pages of both the American and the European press.

- The Republican majority in the House of Representatives choosing to conduct a futile investigation (i.e., without the grant of subpoena power to the Democratic minority) into the government's murderously incompetent response to Hurricane Katrina.

- Within a month of the hurricane's making windfall across 250 miles of shoreline, the award of contracts worth $1.5 billion (for clearing away wreckage, supplying temporary shelter) to corporations (among them Halliburton; Bechtel; Kellogg, Brown and Root) notorious for their profiteering (in amounts as high as $10.5 billion) in the ruins of Iraq.

- Suspension of the rule that requires employers to pay the minimum wage to workers hired on to federally financed construction projects; the waiving of clean-air standards for gasoline in all fifty states; government subcontractors excused from the obligation to submit an affirmative-action plan.

- Among the corps of friends and lobbyists seeking favors on behalf of KBR and The Shaw Group, the presence of Joe M. Allbaugh, President Bush's political campaign manager in 2000, appointed director of FEMA in 2001. On departing the agency in 2003, Allbaugh was replaced by Michael Brown, his deputy and college roommate, until 2001 the Judges and Stewards Commissioner for the International Arabian Horse Association.

- The prices for home inspection services reported by the *New York Times* to be ranging from $15 to $81 per home in accordance with the avarice of the inspector; so also the cost of ships and ferries deployed for temporary housing—in some circumstances, $13 million for six months; in other circumstances, $70 million. Carnival Cruise Lines hired to house evacuees and government relief workers on three ships docked in New Orleans at the weekly rate of $1,400 per guest—as opposed to the $499 charged to passengers at sea for a week's tour of the western Caribbean.

- Debris-removal contracts for approximately $1 billion awarded to AshBritt Inc., a Florida corporation happily associated with Haley Barbour, the governor of Mississippi, former chairman of the Republican National Committee.

- On credit cards issued to government agencies bringing relief from the storm, the spending limit raised from $15,000 to $250,000.

- The blocking by the Republican majority in the Senate of a legislative initiative intended to provide health care under Medicaid to all low-income victims of the hurricane. The White House unwilling to provide rent vouchers for poor people made homeless by the storm, suggesting instead that they be herded into trailer parks. The House of Representatives steadfastly refusing to amend its punitive new bankruptcy law to alleviate the degree of loss and suffering forced on the good people of Alabama, Mississippi, and Louisiana.

- During the first eighteen days after Katrina's arrival, the writing of guidelines for storm-related contracts placed in the care of David Safavian, chief of federal-procurement policy for the Office of Management and Budget. His work was rudely interrupted on September 19, when he was arrested on charges of lying to federal investigators about his involvement in an influence-peddling scheme that took place in 2002, while he was serving as chief of staff for the General Services Administration.

- On the night of August 30, and again on the morning of August 31, the Southern Pines Electric Power Association in Taylorsville, Mississippi, in receipt of phone messages from Vice President Dick Cheney's office in Washington that dictated the order of priority for the restoration of the region's electricity—first to a privately owned pipeline, then to public hospitals.

- A full-page ad in the *New York Times* on September 29, jointly funded by the Hurricane Katrina Fraud Task Force and the U.S. Postal Inspection Service, alerting readers to the strong likelihood of disaster-related federal crimes. In 36-point type the message read, "DON'T LET THESE GUYS TAKE IT AWAY."

So obvious was the nature of the work in progress everywhere on the Gulf Coast, so many corporate bagmen taking it away in eighteen-wheel trucks, that some of the Republican pit bosses in Washington worried about the keeping up of respectable appearances. Mike Pence (R., Ind.), a prominent conservative in the House of Representatives, hedged his misgivings in fiscally responsible platitudes about Congress's doing its best to "insure that a catastrophe of nature does not become a catastrophe of debt for our children." Richard Skinner, the inspector general for the Department of Homeland Security, was slightly more forthright in his language: "We are very apprehensive about what we are seeing."

So were the country's newspaper columnists, many of them appalled by the thievery taking place in broad daylight but at a loss for words with which to depict the scale of the operation or express the magnitude of their disgust. I appreciate the difficulty; when describing the proceedings in Washington over the last five years, I've often been forced back on the vocabulary of historical precedent—masked outlaws holding up stagecoaches on the old Santa Fe Trail, nineteenth-century robber barons rigging the New York stock market, the hit men loyal to Dutch Schultz or Al Capone.

On further acquaintance with the modus operandi of the Bush Administration, I've come to think that the attributions of a competent criminal intelligence miss the point. They give credit where no credit is due, and they fail to account for both the increasingly evident childishness of American culture and the corollary attitudes of entitlement that over the last thirty years have infected ever-larger sectors of the country's equestrian class. President Bush and his friends bear comparison not to Jesse James or Commodore Vanderbilt but to a clique of spoiled trust-fund kids. Certain of their superiority by virtue of their wealth (whether derived from corporate salary, family inheritance, or a sweetheart real estate investment), they fit the profile of wised-up teenagers who don't want to hear it from anybody telling them what to do—which shoes to wear, how to behave in a dance club, when to speak to the caddie or the French ambassador, why it might not be a good idea to wreck the Social Security system, redirect the flow of the Missouri River, or invade Iraq. Smug in their cynicism, proud of their selfishness, pre-Copernican in the sense that they know it is the sun that revolves around them, not they who revolve around the sun, fortune's children interpret corrections as insult, amendments as impertinence—old news, uncool.

The attitude shows plainly in nearly every expression that wanders across the President's schoolboy face—the sly smirk, the cute smile, the petulant frown. At home on the range with his chainsaw in Crawford, Texas, he looks to be making a guest appearance for Paris Hilton on *The Simple Life*; at a White House podium threatening Arab terrorists or standing tall in his opposition to universal health insurance, he strikes the pose of a rich boy anarchist wishing to frighten the faculty at Yale. Three years ago on Earth Day the news photographs showed the President setting off into the forest with an axe over his shoulder, glancing back at the camera with a hint of malicious mischief, as if to say, "You liberal media guys think that the environment is sacred? Let me show you how we deal with trees."

It is with acts of vandalism that juvenile delinquents proclaim their manhood, and what else is the Bush Administration's record over the last five years if not a testimony to its talent for breaking things—the destruction of Afghanistan and Iraq, the loss of respect for America nearly everywhere in the world, a $236 billion budget surplus in 2000 scrapped for a $412 billion deficit in 2004, the country's economic future consigned to foreign creditors, the ever accelerating dissolution of the American political union into separatist factions of race, religion, gender, and social caste.

Endowed with the same temperament as Billy Carter and Roger Clinton but luckier in the sum of their allowances, President Bush and his clique came to power in Washington with little else in their well-groomed heads except the one big idea central to two generations of Republican speech-making: that government is by definition a homeless shelter. Grover Norquist, president of Americans for Tax Reform and for the last twenty years a rabid voice of radical Republicanism, simplified the political science for the listeners of NPR's *Morning Edition* in May 2001: "I don't want to abolish government. I simply want to reduce it to the size where I can drag it into the bathroom and drown it in the bathtub." Just so, and with language added by Norquist's admiring friend and ally, Karl Rove, the Bush Administration speaks for the kind of people who assign no value even to the idea of government, find no use for such a thing as an American *res publica*. Why should they? What's to learn? Everybody who is anybody in Houston or Palm Beach knows that government is a trailer park for deadbeats who can't afford to hire their own servants, furnish their own police protection, hire cheap Chinese labor, pay their taxes in Bermuda. Government is worth owning for the same reasons that one might own a gambling casino or a brothel, a financially rewarding enterprise staffed with quick-witted pimps and can-do waiters. If government is undeserving of respect, worthless except as a means of money-laundering, then why go to the trouble of hiring well-qualified people to collect the taxes and sit in the chairs? What needs to be done that can't be done by one's college roommate, tennis partner, brother-in-law, former secretary, personal lawyer, or golfing buddy?

Adults spoil the fun. They remind the young heirs that government is a matter of long-term maintenance, a learning how to see, know, and care for other people. The lesson follows from the recognition that the national security doesn't consist in a handsome collection of military uniforms but in the health, well-being, and intelligence of a democratic citizenry. The *jeunesse dorée* don't stoop to maintenance; they find it tedious and boring, not glamorous, apt to take time away from thinking about one's hair. Adults also give offense by not picking up on the importance of teenage loyalties (in the club or out, with us or against us); also by reason of their sometimes trying to tell the truth, which in the Bush Administration is a cause for summary dismissal—former Secretary of the Treasury Paul O'Neill cashiered for having had the effrontery to inform Mr. Bush that money doesn't grow on trees, General Eric K. Shinseki, former chief of staff of the U.S. Army, promptly retired because he told Secretary of Defense Donald Rumsfeld that his video-game war in Iraq would require the participation of several hundred thousand American troops, Bunnatine Greenhouse, competition advocate for the U.S. Army Corp of Engineers, reduced in rank for saying that the KBR oil swindle in Iraq was "the most blatant improper contract abuse I have witnessed in the course of my professional career."

Understand "government" as a synonym for "adult," and what

we have now in Washington is the sovereignty of the state in the careless and resentful grasp of teenage anarchists. The historical precedents are legion, among them the reign of the adolescent Roman Emperor Nero, but maybe I'm unduly pessimistic, and possibly what we have before us is the dawn of a new and golden age. If so, at least some of the credit is deserved by all the good people in the fashion, news, banking, and entertainment industries who have made America great. If Vice President Cheney and his business associates don't know how to think or read, they owe their peace of mind to an educational system that teaches by television clip and film montage; if President Bush and his companions in arms delight in all things shallow, derivative, and dumb, they take their sense of ease and comfort from the assurances of a consumer market and a popular culture that place a high value on those qualities. Who can say that the President doesn't embody the American dream come true?

December 2005

Exit Strategies

It is not obligatory for a generation to have great men.
—José Ortega y Gasset

*A*s it becomes increasingly evident that the war in Iraq isn't likely to lead to a happy, Hollywood ending, an ever larger number of its once-upon-a-time champions—cost-conscious Republicans as well as conscience-stricken Democrats—have begun to suffer increasingly severe shortages of memory. On their better days they can remember that Iraq is a faraway Arab country, famous for its mosques and palm trees, but when asked why Baghdad is burning, or how it has come to pass that 2,096 American soldiers are no longer reporting for work on what in the winter of 2003 was imagined as a movie set, they become anxious and forgetful. Last fall's sudden rise in newly discovered cases of amnesia coincided with the season's news reports about the Bush Administration's having set up the invasion of Iraq behind a screen of flag-waving lies—the CIA misinforming the Pentagon, the Pentagon falsifying its dispatches to the State Department, the White

House gulling the Congress, Congress running a shell game on itself.

Given the multiple choice of reasons for not knowing what was what (then, now, preferably never), the convenient losses of memory also could be construed as symptoms of a too trusting faith in the goodness of one's fellow man, and during the months of October and November the Washington talk-show circuit was loud with displays of indignant surprise and wet with the tears of betrayal. Everybody a blameless dupe—misled, played for a sucker, sold down the rivers of deception—and therefore nobody responsible for the casualty lists and the dead dream of empire. Nothing wrong with anybody's character or motives, of course; nobody here in the television studio or the House of Representatives except a patriotic assembly of loyal Americans overwhelmed by a massive systems failure, which is a technical problem, not a sign of bad faith or a proof of blind stupidity. The lights went out; the secretaries forgot to put the truth in the water.

Some of the stories deserved accompaniment for solo violin, others were best understood as acts of contrition on loan from the National Cathedral, but all of them clung to the skirts of the same script. Thus Brent Scowcroft, former national security adviser to the first President George Bush, opposed to the theory of the Second Gulf War, appalled by Vice President Dick Cheney's office deploying against enemies both foreign and domestic the strategies of forward deterrence and preemptive strike, telling a writer for *The New Yorker,* "I consider Cheney a good friend—I've known him for thirty years. But Dick Cheney I don't know anymore."

Or Senator John Kerry, erstwhile presidential candidate who in October 2002 had endorsed the glorious march on Baghdad, speaking to an audience at Georgetown University on October 26:

> I regret that we were not given the truth; as I said more than
> a year ago, knowing what we know now, I would not have

gone to war in Iraq. And knowing now the full measure of the
Bush Administration's duplicity and incompetence, I doubt
there are many members of Congress who would give them
the authority they have abused so badly. I know I would not.

Or the bewildered journalist George Packer, publishing a 467-
page book, *The Assassins' Gate*, in which he deconstructs every
policy initiative and bureaucratic maneuver preliminary to the
American assault on Baghdad, but finding at the end of his labors
that he can't answer the question "Why did the United States invade Iraq? It still isn't possible to be sure—and this remains the
most remarkable thing about the Iraq War." Unwilling or unable to
guess at what he calls "the real motives of the Bush administration," Packer declares himself a victim of his own idealism, decides
that "Iraq is the *Rashomon* of wars," and concludes that the reason
for it "has something to do with September 11."

By the second week in October no C-SPAN camera lacked for a
talking head pleading its inability to distinguish fact from fiction.
So many people had been so wickedly deceived that even the editor
of the *New York Times* had been lost in the fog of disinformation,
failing to notice that Judith Miller, a star reporter for his own newspaper, also was operating as a conduit for government propaganda.
Before the last leaves of autumn had fallen from the trees on Capitol Hill it had become hard to judge which of the testimonials was
the most endearing or instructive. The committees of liberal conscience in town praised Packer's soft-headedness, approved Scowcroft's geopolitical modesty, admired the trembling of Kerry's chin,
but the gold medal for moral awakening they awarded to Colonel
Lawrence Wilkerson, a retired Army officer who from 2002 to
2005 had served as chief of staff to Secretary of State Colin Powell

and who appeared at the podium of the New America Foundation on October 19 to say that during his long career in government (as a staff officer and as a scholar) he had studied the twistings, flummoxings, "aberrations," "bastardizations," "perturbations," apt to occur at the highest echelons of power, but never had he seen anything worse than what he had seen in his years with the Bush Administration. "What I saw was a cabal between the vice president of the United States, Richard Cheney, and the secretary of defense, Donald Rumsfeld, on critical issues that made decisions that the bureaucracy did not know were being made."

The colonel's reference to "a cabal"—a daring word, daringly borrowed from the library of antiwar dissent—earned him a moment in the sun of the *New York Times*' op-ed page (as did his saying, of Undersecretary of Defense Douglas Feith, "Seldom in my life have I met a dumber man"), but the columnists who set him up with the laurel leaves (noble teller of truth to the stone face of power) apparently didn't read the full text, which might have curbed their enthusiasm. The document is remarkable for its pedantry, its presumptions of virtue, its childishness. Proud of his postings as a teacher of military science at both the Naval and Marine war colleges, the colonel fancies himself a sage, but, like Packer, whose book he praises as a Boy Scout guide into the wilderness of bureaucratic dysfunction, he doesn't know why the United States declared war on Iraq. The plan was unintelligible, the objective a mystery. Yes, something criminal probably was afoot in the "Oval Office cabal," but the colonel doesn't care to know the details. Not because he doesn't deplore the abuses of government power but because good American boys don't consort with cabals, don't go into the woods where the wild things are, don't fool around with their sisters. More inclined to preserve his own state of grace than to mess around with snakes, and as unwilling as Packer to think for himself, the colonel devotes the bulk of his text to statements of high-minded bureaucratic principle supported by innovative suggestions for more effective corporate management:

> The complexity of the crises that confront governments today are just unprecedented. . . . You simply cannot deal with all the challenges that government has to deal with, meet all the demands that government has to meet in the modern age, in the twenty-first century, without admitting that it is hugely complex.
>
> And if something comes along that is truly serious, truly serious, something like a nuclear weapon going off in a major American city, or something like a major pandemic, you are going to see the ineptitude of this government in a way that will take you back to the Declaration of Independence. . . . [R]ead in there what they [the Framers] say about the necessity of the people to throw off tyranny or to throw off ineptitude or to throw off that which is not doing what the people want it to do. And you're talking about the potential for, I think, real dangerous times if we don't get our act together.
>
> I really think we have to protect ourselves against institutional imperfections, and in particular we have to protect ourselves against the institutions of humans and the imperfections that we bring.
>
> I like to use the word gracelessness, and I use that word because grace is something we have lost in the modern world. It's a very important product.
>
> We can't leave Iraq. We simply can't. . . . But we're there, we've done it, and we cannot leave. I would submit to you that if we leave precipitously or we leave in a way that doesn't leave something there we can trust, if we do that, we will mobilize the nation, put 5 million men and women under arms

and go back and take the Middle East within a decade. That's what we'll have to do. So why not get it right now?

[T]he world is essentially fractious today and failed states are the future, not the past, and we are the proprietor. It is our obligation and our responsibility in some cases to be a good proprietor. In other cases we have to be more realistic.

You never know what you are going to need on the battlefield, so you'd better have six of them. Five of them won't show up, four of them won't be able to communicate, and I could go on. But you need overlap, you need redundancy. You need, as Powell used to say, "decisive force." You'd better have ten cases of water where you think you'd need one. You'd better have 15 million MREs where you think you need only a million because you never know in a crisis, and the best way to be prepared is to have lots more than you think you're going to need or want.

It might also be prudent to have on hand a surplus of intelligence, but if the tone and quality of the colonel's thought is representative of what passes for wisdom in the head of the American government, where then is the hope of confronting the "hugely complex" challenges of the twenty-first century with anything other than a childish belief in magic? After reading the transcript of the presentation to the New America Foundation, I watched the rerun of the television broadcast, which, unhappily, didn't correct the impression of a charismatic Christian speaking in tongues. I could see that the colonel was probably a very nice man, earnest and well-intentioned, proceeding diligently from power point to power point, here to help and not to hurt, but so lost in the ritual language of bureaucratic abstraction that although he presumably knew what he was talking about, he undoubtedly didn't know that what he was talking about wasn't worth knowing.

More than once he repeated a dire warning with the emphasis of implied exclamation points ("problems are brewing! problems are brewing! . . . My army right now is truly in bad shape—truly in bad shape!"), but when something goes wrong in America it isn't because anybody in government means to lie, cheat, steal, commit murder, or otherwise do harm. How could they? They're Americans and therefore good. It's never the people who are at fault; it's because the system is "dysfunctional," because the intelligence agencies "don't share," "never talk to each other," don't grasp the fact that everybody's "got to work together . . . under leadership they trust and leadership that on basic issues they agree with. . . ."

It wasn't until I'd read through the colonel's *cri de coeur* for a second and third time that I began to understand how it could happen that so many of Washington's nominally well-informed politicians and journalists suffered so massive an intelligence failure prior to the invasion of Iraq, or why the same cloud of unknowing hadn't descended on the conversation in New York. By late January 2003, six weeks before the bombs fell on Baghdad, the Bush Administration's stated reasons for going to war already had been shown to be fraudulent, and despite the news media's doing their patriotic best not to notice what was wrong with the sales pitch, the swindle was a matter of public record—Andrew Card, the President's chief of staff, had suggested to the *New York Times* in September of 2002 that the timing of the assault on Baghdad was mostly a matter of marketing; the U.N. weapons inspectors during the autumn of that year had made numerous journeys to Iraq, finding no instruments of mass destruction; Saddam Hussein's supposed connection to Al Qaeda was clearly illusory; Vice President Cheney's intelligence operatives and those under contract to the CIA were quarreling openly in the newspapers about the data gathered from sources dubious and self-serving, reliable only to the extent that they could be trusted to say what they had been paid to say.

The available facts were consistent with what was known at the time about the Bush Administration's will to power and with what could be reasonably inferred about its commercial motive and imperial intent, the postulates easily enough obtained merely by numbering the false statements in any one of President Bush's speeches, or simply by watching the Pentagon press briefings at which Secretary of Defense Donald Rumsfeld's attitude implied that the waging of war in Central Asia really wasn't much different than the sending of Air Force jets to perform a flyby overhead the Superbowl. Nobody needed access to privileged gossip or a talent for interpreting aerial reconnaissance photographs to know that the President wanted a war in Iraq, that he possessed the means to get what he wanted (a cowed legislature, an accommodating press, an inert electorate), and that it didn't matter what reasons were given for the blitzkrieg—exporting democracy, winning World Wars III and IV, saving Israel, protecting America, bringing the Christian faith to heathen Islam, etc.—as long as they came wrapped with the ribbon of the American flag.

Such at least was the general understanding on the part of the many people (by some estimates at least 800,000 people) who on February 15, 2003, staged street demonstrations in 150 American cities as a way of voicing their skepticism. Maybe they didn't know whether it was the Euphrates or the Tigris River that flowed through Baghdad, but they could recognize the difference between the truth and its expedient equivalents.

The capacity to notice the difference and the willingness to act on the observation presuppose the mind and presence of an adult— i.e., an individual whose character and moral sense is formed by his or her own thought and experience. Washington these days doesn't have much use for adults; they can't be trusted to go along with the program, play well with others, believe what they read in the newspapers. What is wanted is a quorum of dutiful children, who know that skepticism is wicked and credulity a virtue that also stands and serves as job requirement for their successful rising in the ranks of the government and media bureaucracies. Like the anxious courtiers in feathered hats who once decorated the throne rooms of old Europe, they fit their convictions to the circumstance, borrow their sense and sensibility from the consensus present in the school dormitory or the Senate conference committee, in this year's color scheme or last week's opinion poll. If from time to time the consensus changes (the war in Iraq is good, the war in Iraq is bad), staff officers as well trained as Colonel Wilkerson in the art of devising exit strategies and politicians as willing as Senator John Kerry to change trains know that the American public would rather comfort a child than pardon a criminal or forgive a fool.

January 2006

The Case for Impeachment

Why We Can No Longer Afford George W. Bush

A country is not only what it does—it is also what it puts up with, what it tolerates.
—Kurt Tucholsky

O n December 18 of last year, Congressman John Conyers Jr. (D., Mich.) introduced into the House of Representatives a resolution inviting it to form "a select committee to investigate the Administration's intent to go to war before congressional authorization, manipulation of pre-war intelligence, encouraging and countenancing torture, retaliating against critics, and to make recommendations regarding grounds for possible impeachment." Although buttressed two days previously by the news of the National Security Agency's illegal surveillance of the American citizenry, the request attracted little or no attention in the press—nothing on television or in the major papers, some scattered applause from the left-wing blogs, heavy sarcasm on the websites flying the flags of the militant right. The nearly complete silence raised the question as to what it was the congressman had in mind, and to whom did he think he was speaking? In time of war few propositions would seem as futile as the attempt to impeach a president whose political party controls the Congress; as the ranking member of the House Judiciary Committee stationed on Capitol Hill for the last forty years, Representative Conyers presumably knew that to expect the Republican caucus in the House to take note of his invitation, much less arm it with the power of subpoena, was to expect a miracle of democratic transformation and rebirth not unlike the one looked for by President Bush under the prayer rugs in Baghdad. Unless the congressman intended some sort of symbolic gesture, self-serving and harmless, what did he hope to prove or to gain? He answered the question in early January, on the phone from Detroit during the congressional winter recess.

"To take away the excuse," he said, "that we didn't know." So that two or four or ten years from now, if somebody should ask, "Where were you, Conyers, and where was the United States Congress?" when the Bush Administration declared the Constitution inoperative and revoked the license of parliamentary government, none of the company now present can plead ignorance or temporary insanity, can say that "somehow it escaped our notice" that the President was setting himself up as a supreme leader exempt from the rule of law.

A reason with which it was hard to argue but one that didn't account for the congressman's impatience. Why not wait for a showing of supportive public opinion, delay the motion to impeach until after next November's elections? Assuming that further investigation of the President's addiction to the uses of domestic espionage finds him nullifying the Fourth Amendment rights of a large number of his fellow Americans, the Democrats possibly could come up with enough votes, their own and a quorum of disenchanted Republicans, to send the man home to Texas. Conyers said:

"I don't think enough people know how much damage this administration can do to their civil liberties in a very short time. What would you have me do? Grumble and complain? Make cynical jokes? Throw up my hands and say that under the circumstances nothing can be done? At least I can muster the facts, establish a record, tell the story that ought to be front-page news."

Which turned out to be the purpose of his House Resolution 635—not a high-minded tilting at windmills but the production of a report, 182 pages, 1,022 footnotes, assembled by Conyers's staff during the six months prior to its presentation to Congress, that describes the Bush Administration's invasion of Iraq as the perpetration of a crime against the American people. It is a fair description. Drawing on evidence furnished over the last four years by a sizable crowd of credible witnesses—government officials both extant and former, journalists, military officers, politicians, diplomats domestic and foreign—the authors of the report find a conspiracy to commit fraud, the administration talking out of all sides of its lying mouth, secretly planning a frivolous and unnecessary war while at the same time pretending in its public statements that nothing was further from the truth.[1] The result has proved tragic, but on reading through the report's corroborating testimony I sometimes could counter its inducements to mute rage with the thought that if the would-be lords of the flies weren't in the business of killing people, they would be seen as a troupe of off-Broadway comedians in a third-rate theater of the absurd. Entitled "The Constitution in Crisis; The Downing Street Minutes and Deception, Manipulation, Torture, Retribution, and Coverups in the Iraq War," the Conyers report examines the administration's chronic abuse of power from more angles than can be explored within the

---

[1] *The report borrows from hundreds of open sources that have become a matter of public record—newspaper accounts, television broadcasts* (Frontline, Meet the Press, Larry King Live, 60 Minutes, *etc.*), *magazine articles* (*in* The New Yorker, Vanity Fair, The New York Review of Books), *sworn testimony in both the Senate and House of Representatives, books written by, among others, Bob Woodward, George Packer, Richard A. Clarke, James Mann, Mark Danner, Seymour Hersh, David Corn, James Bamford, Hans Blix, James Risen, Ron Suskind, Joseph Wilson. As the congressman had said, "Everything in plain sight; it isn't as if we don't or didn't know."*

compass of a single essay. The nature of the administration's criminal DNA and modus operandi, however, shows up in a singularly vivid specimen of its characteristic dishonesty.

That President George W. Bush comes to power with the intention of invading Iraq is a fact not open to dispute. Pleased with the image of himself as a military hero, and having spoken, more than once, about seeking revenge on Saddam Hussein for the tyrant's alleged attempt to "kill my Dad," he appoints to high office in his administration a cadre of warrior intellectuals, chief among them Secretary of Defense Donald Rumsfeld, known to be eager for the glories of imperial conquest.[2] At the first meeting of the new National Security Council on January 30, 2001, most of the people in the room discuss the possibility of preemptive blitzkrieg against Baghdad.[3] In March the Pentagon circulates a document entitled "Foreign Suitors for Iraqi Oil Field Contracts"; the supporting maps

[2] *In January of 1998 the neoconservative Washington think tank The Project for the New American Century (which counts among its founding members Dick Cheney) sent a letter to Bill Clinton demanding "the removal of Saddam Hussein's regime from power" with a strong-minded "willingness to undertake military action." Together with Rumsfeld, six of the other seventeen signatories became members of the Bush's first administration—Elliott Abrams (now George W. Bush's deputy national security advisor), Richard Armitage (deputy secretary of state from 2001 to 2005), John Bolton (now U.S. ambassador to the U.N.), Richard Perle (chairman of the Defense Policy Board from 2001 to 2003), Paul Wolfowitz (deputy secretary of defense from 2001 to 2005), Robert Zoellick (now deputy secretary of state). President Clinton responded to the request by signing the Iraq Liberation Act, for which Congress appropriated $97 million for various clandestine operations inside the borders of Iraq. Two years later, in September 2000, The Project for the New American Century issued a document noting that the "unresolved conflict with Iraq provides the immediate justification" for the presence of the substantial American force in the Persian Gulf.*
[3] *In a subsequent interview on* 60 Minutes, *Paul O'Neill, present in the meeting as the newly appointed secretary of the treasury, remembered being surprised by the degree of certainty: "From the very beginning, there was a conviction that Saddam Hussein was a bad person and that he needed to go. . . . It was all about finding a way to do it."*

indicate the properties of interest to various European governments and American corporations. Six months later, early in the afternoon of September 11, the smoke still rising from the Pentagon's western facade, Secretary Rumsfeld tells his staff to fetch intelligence briefings (the "best info fast . . . go massive; sweep it all up; things related and not") that will justify an attack on Iraq. By chance the next day in the White House basement, Richard A. Clarke, national coordinator for security and counterterrorism, encounters President Bush, who tells him to "see if Saddam did this." Nine days later, at a private dinner upstairs in the White House, the President informs his guest, the British prime minister, Tony Blair, that "when we have dealt with Afghanistan, we must come back to Iraq."

By November 13, 2001, the Taliban have been rousted out of Kabul in Afghanistan, but our intelligence agencies have yet to discover proofs of Saddam Hussein's acquaintance with Al Qaeda.[4] President Bush isn't convinced. On November 21, at the end of a National Security Council meeting, he says to Secretary Rumsfeld, "What have you got in terms of plans for Iraq? . . . I want you to get on it. I want you to keep it secret."

The Conyers report doesn't return to the President's focus on Iraq until March 2002, when it finds him peering into the office of Condoleezza Rice, the national security advisor, to say, "Fuck Saddam. We're taking him out." At a Senate Republican Policy lunch that same month on Capitol Hill, Vice President Dick Cheney informs the assembled company that it is no longer a question of if the United States will attack Iraq, it's only a question of when. The vice president doesn't bring up the question of why, the answer to which is a work in progress. By now the administration knows, or at least has reason to know, that Saddam Hussein had nothing to do with the

---

[4] *As early as September 20, Douglas Feith, undersecretary of defense for policy, drafted a memo suggesting that in retaliation for the September 11 attacks the United States should consider hitting terrorists outside the Middle East in the initial offensive, or perhaps deliberately selecting a "non-Al Qaeda target like Iraq."*

9/11 attacks on New York and Washington, that Iraq doesn't possess weapons of mass destruction sufficiently ominous to warrant concern, that the regime destined to be changed poses no imminent threat, certainly not to the United States, probably not to any country defended by more than four batteries of light artillery. Such at least is the conclusion of the British intelligence agencies that can find no credible evidence to support the theory of Saddam's connection to Al Qaeda or international terrorism; "even the best survey of WMD programs will not show much advance in recent years on the nuclear, missile and CW/BW weapons fronts . . ." A series of notes and memoranda passing back and forth between the British Cabinet Office in London and its correspondents in Washington during the spring and summer of 2001 address the problem of inventing a pretext for a war so fondly desired by the Bush Administration that Sir Richard Dearlove, head of Britain's MI-6, finds the interested parties in Washington fixing "the intelligence and the facts . . . around the policy." The American enthusiasm for regime change, "undimmed" in the mind of Condoleezza Rice, presents complications.

Although Blair has told Bush, probably in the autumn of 2001, that Britain will join the American military putsch in Iraq, he needs "legal justification" for the maneuver—something noble and inspiring to say to Parliament and the British public. No justification "currently exists." Neither Britain nor the United States is being attacked by Iraq, which eliminates the excuse of self-defense; nor is the Iraqi government currently sponsoring a program of genocide. Which leaves as the only option the "wrong-footing" of Saddam. If under the auspices of the United Nations he can be presented with an ultimatum requiring him to show that Iraq possesses weapons that don't exist, his refusal to comply can be taken as proof that he does, in fact, possess such weapons.[5]

---

[5] *Abstracts of the notes and memoranda, known collectively as "The Downing Street Minutes," were published in the* Sunday Times *(London) in May 2005; their authenticity was undisputed by the British government.*

Over the next few months, while the British government continues to look for ways to "wrong-foot" Saddam and suborn the U.N., various operatives loyal to Vice President Cheney and Secretary Rumsfeld bend to the task of fixing the facts, distributing alms to dubious Iraqi informants in return for map coordinates of Saddam's monstrous weapons, proofs of stored poisons, of mobile chemical laboratories, of unmanned vehicles capable of bringing missiles to Jerusalem.[6]

By early August the Bush Administration has sufficient confidence in its doomsday story to sell it to the American public. Instructed to come up with awesome text and shocking images, the White House Iraq Group hits upon the phrase "mushroom cloud" and prepares a White Paper describing the "grave and gathering danger" posed by Iraq's nuclear arsenal.[7] The objective is three-fold—to magnify the fear of Saddam Hussein, to present President Bush as the Christian savior of the American people, a man of conscience who never in life would lead the country into an unjust war, and to provide a platform of star-spangled patriotism for Republican candidates in the November congressional elections.[8]

The ad campaign rolls out on September 7, when Britain's Tony Blair stands in front of the television cameras at Camp David,

---

[6] The work didn't go unnoticed by people in the CIA, the Pentagon, and the State Department accustomed to making distinctions between a well-dressed rumor and a naked lie. In the spring of 2004, talking to a reporter from Vanity Fair, Greg Thielmann, the State Department officer responsible for assessing the threats of nuclear proliferation, said, "The American public was seriously misled. The Administration twisted, distorted and simplified intelligence in a way that led Americans to seriously misunderstand the nature of the Iraq threat. I'm not sure I can think of a worse act against the people in a democracy than a President distorting critical classified information."

[7] The Group counted among its copywriters Karl Rove, senior political strategist, Andrew Card, White House chief of staff, National Security Advisor Condoleezza Rice, and Lewis "Scooter" Libby, Dick Cheney's chief of staff.

[8] Card later told the New York Times that "from a marketing point of view . . . you don't introduce new products in August."

Maryland, with President Bush to say that a new report from the International Atomic Energy Agency shows new activity at Iraq's nuclear weapons sites.[9] On September 8, National Security Advisor Rice appears on *Late Edition with Wolf Blitzer,* to picture a mushroom cloud in America's future and Defense Secretary Rumsfeld, on *Face the Nation,* invites Bob Schieffer to "imagine a September 11 with weapons of mass destruction." On the same day, Vice President Cheney shows up on *Meet the Press* to assure Tim Russert that "first of all, no decision's been made yet to launch a military operation." The President stays on both messages, informing reporters gathered at an Oval Office photo op on September 25 that when discussing the War on Terror, "You can't distinguish between Al Qaeda and Saddam," and then on October 1, after meeting with members of Congress, "Of course, I haven't made up my mind if we're going to war with Iraq."[10]

The autumn sales promotion reaches its crescendo on October 7, when the President, speaking to a live television audience from the Cincinnati Museum Center, pulls all the dead rabbits out of Karl Rove's magic hat—Saddam possessed of "horrible poisons and diseases and gasses and atomic weapons . . . we know that Iraq and Al Qaeda have had high-level contacts that go back a decade. . . . America must not ignore the threat gathering against us. Facing clear evidence of peril, we cannot wait for the final proof—the smoking gun—that could come in the form of a mushroom cloud . . ."

Four days later in Washington, on the evening of October 11, Congress passes HJ Resolution 114, granting President Bush the

---

[9] *Bush confirms Blair's statement, saying, "I don't know what more evidence we need." In Vienna the day before the IAEA issued a statement saying that there was no report.*

[10] *Collaborating in what was a team effort between March 2002 and March 2004, various high-ranking administration officials made 237 false or misleading statements (55 of them from President Bush himself) connecting Saddam to Al Qaeda, exaggerating Iraq's biological and chemical weapons capabilities, misrepresenting Iraq's nuclear activities.*

power to "use the Armed Forces of the United States as he determines to be necessary and appropriate in order to defend the national security of the United States against the continuing threat posed by Iraq." The grant excuses the President from the obligation to consult Congress on a formal declaration of war; its preamble reiterates the administration's fantastic assertions about "Iraq's demonstrated capability and willingness to use weapons of mass destruction, the risk that the current Iraqi regime will either employ those weapons to launch a surprise attack against the United States or its Armed Forces or provide them to international terrorists who would do so, and the extreme magnitude of harm that would result to the United States and its citizens from such an attack, combine to justify action by the United States to defend itself . . ." Passage of the resolution, by a vote of 77 to 23 in the Senate, certifies as true what is known to be false and thus enlists Congress as an accomplice in the systematic perpetration of a criminal fraud.

Its war policy thus firmly established in all the major media markets, the Bush Administration over the span of the next five months holds fast to the policy of deceiving itself as well as the American people and the Congress. While the Pentagon assembles its forces for Operation Iraqi Freedom on March 20, 2003, President Bush continues to present himself as the victim of outrageous circumstance. In answer to a reporter's question at a White House press conference, he says, "You said we're headed to war in Iraq—I don't know why you say that. I hope we're not headed to war in Iraq. I'm the person who gets to decide, not you." Vice President Cheney tells Hans Blix, the chief U.N. weapons inspector, that unless his scouts soon find Saddam's WMD in Iraq, the United States "will not hesitate to discredit inspections in favor of disarmament"; meanwhile, having come to believe the lies stuffed into his mouth by informants paid to do just that, he appears on *Meet the Press* to say, "My belief is we will, in fact, be greeted as liberators." President

Bush in his State of the Union Address on January 28 falsely informs Congress that Saddam has been trying to buy enriched uranium in Africa; at the U.N. Security Council on February 5, Secretary of State Colin Powell conjures up the "sinister nexus between Iraq and the Al Qaeda terrorist network."[11]

The Conyers report doesn't lack for further instances of the administration's misconduct, all of them noted in the press over the last three years—misuse of government funds, violation of the Geneva Conventions, holding without trial and subjecting to torture individuals arbitrarily designated as "enemy combatants," etc.—but conspiracy to commit fraud would seem reason enough to warrant the President's impeachment.[12] Before reading the report, I wouldn't have expected to find myself thinking that such a course of action was either likely or possible; after reading the report, I don't know why we would run the risk of not impeaching the man. We have before us in the White House a thief who steals the country's good name and reputation for his private interest and personal use; a liar who seeks to instill in the American people a state of fear; a televangelist who engages the United States in a never-ending crusade against all the world's evil, a wastrel who squanders a vast sum of the nation's wealth on what turns out to be a recruiting

---

[11] *Powell occasionally complained about the falsehoods the administration obliged him to tell; a few days before delivering the U.N. speech he mentioned his unhappiness to Cheney, who told him, "Your poll numbers are in the seventies, you can afford to lose a few points."*

[12] *The legal precedent for finding a conspiracy to commit fraud against the United States rests on the Supreme Court ruling* Hammerschmidt v. United States, *which upholds the charge against individuals who obstruct lawful government functions "by deceit, craft or trickery, or at least by means that are dishonest. It is not necessary that the government shall be subjected to property or pecuniary loss by the fraud, but only that its legitimate official action and purpose shall be defeated by misrepresentation, chicane or the over-reaching of those charged with carrying out the government intention."*

drive certain to multiply the host of our enemies. In a word, a criminal—known to be armed and shown to be dangerous.[13] Under the three-strike rule available to the courts in California, judges sentence people to life in jail for having stolen from Wal-Mart a set of golf clubs or a child's tricycle. Who then calls strikes on President Bush, and how many more does he get before being sent down on waivers to one of the Texas Prison Leagues?

Last December when Representative Conyers introduced House Resolution 635—together with two ancillary resolutions censuring President Bush and Vice President Cheney, each of them for "failing to respond to requests for information" about the origins of the Iraq war—the word "impeach" was still regarded as excessive hyperbole, the unmentionable "I-word," not to be seen or heard in the circles of proper Washington opinion. A month later and the word was in general circulation, if not with reference to the misbegotten killing in Iraq at least in the context of the President's directing the NSA to monitor, if necessary without first obtaining a court order, any and all telephone and email traffic, in, out, or around the United States, that might turn up a reference (ideological, operational, metaphorical, coincidental, or conversational) to an act (factual or fictional, past, present, or future) of terrorism. So wide a spreading of government flypaper obviously was bound to collect (by accident, if not by design) a great deal of information about a great many American citizens under the impression that their words were their own. The President's directive, a felony under the 1978 Foreign Intelligence Surveillance Act (punishable by five years in prison and a $10,000 fine), also nullified

---

[13] As of January 17, 2006, the rap sheet listed 2,229 American military dead in Iraq together with an unknown number of Iraqi civilians; what looks to be the sum of $1 trillion, by some estimates $2 trillion, already committed to The Project for the New American Century's real estate development in the Mesopotamian desert. Better reasons to impeach a president than the one pressed into service against Bill Clinton, whose penis was known to be aimless and shown to be harmless.

the Fourth Amendment's guarantee of protection against unreasonable search and seizure.

As was to be expected, the first stirrings of objection appeared in the leftwing blogs, tens of thousands of citizens posting violently worded dissents on impeachbush.org, also at afterdowningstreet.org and impeachcentral.com. On December 19, two days after President Bush affirmed his commander in chief's right to affix wire taps whenever and wherever he so chose, Senator Robert Byrd (D., W.Va.) issued a statement saying that "we are a nation of laws and not men. . . . I defy the Administration to show me where in the Foreign Intelligence Surveillance Act, or the U.S. Constitution, they are allowed to steal into the lives of innocent Americans and spy. . . . These astounding revelations about the bending and contorting of the Constitution to justify a grasping irresponsible administration under the banner of 'national security' are an outrage. . . ." Representative John Lewis (D., Ga.), told a radio interviewer in Atlanta that George W. Bush "is not a king, he is President," going on to say that if the chance presented itself he would sign a bill of impeachment. John Dean, White House counsel in the Nixon Administration and a man familiar with the arts of bugging phones and obstructing justice, observed, in conversation with Senator Barbara Boxer (D., Calif.), that "Bush is the first president to admit to an impeachable offence."[14]

In the conservative quarters of opinion the objections were equally vigorous. On December 19, Norman Ornstein, a scholar at the American Enterprise Institute allied with the Republican

[14] *By the second week in January a Zogby poll showed a majority of Americans (52–43 percent) favoring impeachment of the President if he were to be found entangled in the coils of illegal surveillance; a resolution to that effect had been carried with rousing applause by the city council in Arcata, California; seven other members of the House of Representatives had come forward to co-sponsor Conyers's Resolution 635—Lois Capps (D., Calif.), Sheila Jackson-Lee (D., Tex.), Zoe Lofgren (D., Calif.), Donald Payne (D., N.J.), Charles Rangel (D., N.Y.), Maxine Waters (D., Calif.), and Lynn Woolsey (D., Calif.).*

doctrine of small governments supporting large armies, sat for an interview on a Washington public radio station known for its leftist bias and liberal audience. "I think that if we're going to be intellectually honest here," he said, "this really is the kind of thing that Alexander Hamilton was referring to when impeachment was discussed." The next day, writing in the *Washington Times,* Bruce Fein, former associate deputy attorney general under President Ronald Reagan, said of President Bush that he "presents a clear and present danger to the rule of law"; he later extended the thought by saying that if the President "maintains this disregard or contempt for the coordinate branches of government, it's that conception of an omnipotent presidency that makes the occupant a dangerous person."

By the third week in January the word "impeachment" had been mentioned often enough in polite company to achieve a presence in the vocabulary of Senator Arlen Specter (R., Pa.), chairman of the Senate Judiciary Committee. Appearing on ABC's *This Week* on January 15, Specter said that he planned a February hearing on the matter of the warrantless domestic spying program, but even if it were to be found that the President exceeded his authority, impeachment was not the appropriate rebuke. Pressed on the point by George Stephanopoulos, Specter said: "Impeachment is a remedy. After impeachment, you could have a criminal prosecution, but the principal remedy, George, under our society is to pay a political price."[15]

Would that it were so. The Bush Administration doesn't believe in the theory of parliamentary government, much less in the notion of

[15] *Together with Specter, two prominent Republican senators called for congressional hearings on the matter of the NSA's spying program, Chuck Hagel (Nebr.) and Olympia Snowe (Maine); seven others expressed serious misgivings—Richard Lugar (Ind.), Susan Collins (Maine), Sam Brownback (Kans.), John Sununu (N.H.), Larry Craig (Idaho), Lindsey Graham (S.C.), John McCain (Ariz.).*

paying political prices. Its agents prefer the more frugal and efficient practice of stealing elections, gagging the voice of the Democratic minority in Congress, slandering people who presume to
doubt either its wisdom or its virtue, conducting the business of
government behind closed doors, alone with its Bibles and its pet
Bismarcks in soundproof rooms. On the off chance that any God-
fearing citizen didn't know what to expect, the President clarified
the position in late December and again in early January, responding to what he clearly regarded as annoying questions about his directive to the NSA. Yes, he had told the NSA to take precautions,
had done so more than once, would do so again. That was his job, to
defend the American people; in time of war the Constitution gave
him the right to do as he pleased, so did the act of Congress passed
on September 14, 2001, three days after the loss of the World Trade
Center. The fact that he was compelled to address the subject was
"shameful," impertinent, and unpatriotic on the part of the reporter who inquired about "unchecked executive power" and ascribed to him "some kind of dictatorial position." Such questions
also were dangerous, apt to bring on more terrorist attacks in the
manner of 9/11. The latter point was repeatedly reinforced by Vice
President Cheney, who firmly reminded audiences in New York and
Washington—audiences composed primarily of lobbyists for the
country's media syndicates and weapons manufacturers—that we
live in a dangerous world, demanding a robust executive authority
in the White House to ward off the forces of moral anarchy and social chaos.

"We're at war," the President said on December 19, "we must
protect America's secrets."

No, the country isn't at war, and it's not America's secrets that
the President seeks to protect. The country is threatened by freebooting terrorists unaligned with a foreign government or an enemy army; the secrets are those of the Bush Administration, chief
among them its determination to replace a democratic republic

with something more reliably totalitarian. The fiction of permanent war allows it to seize, in the name of the national security, the instruments of tyranny.

The question posed to the assembly is whether enough people care, and, if so, how do they respond when, in the language of the Declaration of Independence, "a long train of abuses and usurpations pursuing invariably the same Object evinces a design to reduce them under absolute Despotism." Although the abuses and usurpations are self-evident, obvious to anybody who takes the trouble to read the newspapers, the Bush Administration makes no attempt to conceal the Object evinced in the design of its purpose because it counts on the romanticism as well as on the apathy of an American public reluctant to recognize the President of the United States as a felon. Who wants to believe such a thing, much less acknowledge it as a proven fact?

More people than dreamed of in the philosophy of Karl Rove or by the content providers to the major news media. The heavy volume of angry protest on the Internet, reflected in the polls indicating the President's steady decline in the popular esteem, suggests that at least half of the American electorate, in the red states and the blue, knows that the Bush Administration operates without reference to the rule of law, also that the President believes himself somehow divinely ordained, accountable only to Jesus and the oil companies, at liberty to wave what he imagines to be the scepter of the Constitution in whatever ways he deems best. But in the news media they find no strong voice of dissent, in the Democratic Party no concerted effort to form a coherent opposition.

Which places the work of protecting the country's freedoms where it should be placed—with the Congress, more specifically with the Republican members of Congress. What else is it that voters expect the Congress to do if not to look out for their rights as citizens of the United States? The choice presented to the Republican members on the judiciary committee investigating the President's use of electronic surveillance comes down to a matter of deciding whether they will serve their country or their party. I don't envy them the decision; the rewards offered by the party (patronage, campaign contributions, a fat retirement on the payroll of a K Street lobbyist) clearly outweigh those available from the country—congratulatory editorials in obscure newspapers, malicious gossip circulated by Focus on the Family and Fox News, an outpouring of letters and emails from grateful citizens not in positions to do anybody any favors.

To the cameras on ABC's *This Week* Senator Specter said that impeachment was probably not the best remedy for the President's misconduct, and if the February hearings proceed along the lines of most Senate hearings, they will find that the best remedy is no remedy. In rebuttal to testimony arguing that the President's breach of federal law constitutes a crime, other witnesses will say that the Constitution authorizes the President to overrule the law, and after a prolonged mumbling of lawyers about the balance that must be struck between civil liberties on the one hand and the national security on the other, it will be discovered that the balance sometimes shifts to meet the circumstances, which is why, thank God, America is a great country—because it allows for the appearance of meaningful debate while at the same time making sure that the words have no consequences.

If, however the committee rises to the recognition of its constitutional task—that of correcting the imbalances of power that sometimes can foul the machinery of fair and honorable government—it will take to heart the meaning of the word "remedy"—a corrective measure, not a punishment. It isn't the business of the Congress to punish President Bush. Any competent court in the country could arraign the President on charges identical to those brought against the crooked executives at Enron and Tyco International (fraud, misuse of stockholder funds, manipulation of intelligence) and send him off to jail dressed in an orange jump suit. Nor is it the responsibility of Congress to sit in moral judgment; the sermons can be left to the Reverend Pat Robertson and the Yale Divinity School. It is the business of the Congress to prevent the President from doing more damage than he's already done to the people, interests, health, well-being, safety, good name, and reputation of the United States—to cauterize the wound and stem the flows of money, stupidity, and blood.

March 2006

Index